GOD IN THE MACHINE

Ethnography, Theory, Experiment

Series Editors:
Martin Holbraad, Department of Anthropology, University College London
Morten Axel Pedersen, Department of Anthropology, University of Copenhagen
Rane Willerslev, Museum of Cultural History, University of Oslo

In recent years, ethnography has been increasingly recognized as a core method for generating qualitative data within the social sciences and humanities. This series explores a more radical, methodological potential of ethnography: its role as an arena of theoretical experimentation. It includes volumes that call for a rethink- ing of the relationship between ethnography and theory in order to question, and experimentally transform, existing understandings of the contemporary world.

Recent volumes:

Volume 12
GOD IN THE MACHINE: THE GANESH YOURSELF EXPERIMENT
By Emmanuel Grimaud

Volume 11
ENTANGLEMENTS OF THE MAYA UNIVERSE: LIFE, SOCIETY AND COSMOS AMONG THE K'ICHE' OF MOMOSTENANGO
By Alonso Zamora Corona

Volume 10
DREAMING AND THE IMAGINATION: THEORETICAL INTERSECTIONS IN CROSS-CULTURAL PERSPECTIVE
Edited by Matthew D. Newsom

Volume 9
CATTLE POETICS: HOW AESTHETICS SHAPES POLITICS IN MURSILAND, ETHIOPIA
By Jean-Baptiste Eczet

Volume 8
THE CHILDREN OF GREGORIA: DOGME ETHNOGRAPHY OF A MEXICAN FAMILY
By Regnar Kristensen and Claudia Adeath Villamil

Volume 7
GOING TO PENTECOST: AN EXPERIMENTAL APPROACH TO STUDIES IN PENTECOSTALISM
By Annelin Eriksen, Ruy Llera Blanes and Michelle MacCarthy

Volume 6
CUTTING COSMOS: MASCULINITY AND SPECTACULAR EVENTS AMONG THE BUGKALOT
By Henrik Hvenegaard Mikkelsen

Volume 5
VITAL DIPLOMACY: THE RITUAL EVERYDAY ON A DAMMED RIVER IN AMAZONIA
By Chloe Nahum-Claudel

For a full volume listing, please see the series page on our website: https://www.berghahnbooks.com/series/ethnography

GOD IN THE MACHINE
The Ganesh Yourself Experiment

Emmanuel Grimaud

berghahn
NEW YORK • OXFORD
www.berghahnbooks.com

First published in 2026 by
Berghahn Books
www.berghahnbooks.com

Library of Congress Cataloging-in-Publication Data
A C.I.P. cataloging record is available from the Library of Congress
Library of Congress Cataloging in Publication Control Number: 2026006065

British Library Cataloguing in Publication Data
A catalogue record for this book is available from the British Library

EU GPSR Authorized Representative
LOGOS EUROPE, 9 rue Nicolas Poussin, 17000, LA ROCHELLE, France
Email: Contact@logoseurope.eu

ISBN 978-1-83695-527-6 hardback
ISBN 978-1-83695-528-3 epub
ISBN 978-1-83695-529-0 web pdf

https://doi.org/10.3167/9781836955276

The human is a rope, fastened between the animal and the superhuman – a rope across an abyss.

—Friedrich Nietzsche,
Thus Spake Zarathoustra, 1885

Form after form he took on every form, and thus he had a perceivable form.

—Ṛk-Saṃhitā, VI.47.18

CONTENTS

ILLUSTRATIONS

Figure 0.1. Bappa, the tele-operated device conceived by Emmanuel Grimaud and Zaven Paré, was launched in Mumbai during the Ganesh Festival, 2014.

IN THE PLACE OF GOD

The experimental turn in anthropology and its metaphysical stakes

Patrice Maniglier

Here is a book that speaks of God. Here is a book that approaches a subject traditionally associated with the highest speculation, but which does so neither by invoking visionary revelation, nor by deploying subtle reasoning – simply by recounting conversations that the author had previously initiated himself.

Not conversations between great mystics or prominent personalities, but between ordinary people, people without authority – conversations that may recall those overheard at the closing of bars, when the vapours of intoxication lend speculative wings. This is a book where one talks about God. Or perhaps we should say 'god', as one speaks about politics or about love. The dozens of pages of conversations carefully transcribed and reproduced, which make up a substantial part of this book, may seem strange to some – perhaps tedious, at the very least out of place in a work published in a metaphysical collection.

Certainly, the author is not a philosopher by profession; he is an anthropologist. But anthropology is also a theoretical discipline. And the book you are about to read will surely strike you with its apparently under-theorized character and the determination with which it delays any rise to generality, any conceptual act, choosing instead to dwell at length in the delivery of an almost raw material that resembles, in its form, the dialogue scenes of popular novels – or perhaps more accurately, the metaphysical version of the exchanges filmed by Rohmer in *The Tree, the Mayor and the Media Library* – rather than what ethnographers typically call descriptions.

You have to wait until the last two chapters to see the theoretical stakes of the experience fully developed on their own terms. And those stakes are crucial – not only for anthropology but also for metaphysics, and more generally for our understanding of the present moment and the role of the humanities in the time we are living through. It is to provide a few guiding points for those who might get lost in the labyrinth of these conversations – and to help them stay the course until the end of this experiment – that we deemed it useful to write this foreword.

***

'Anthropology is philosophy with people in it.' This sentence by British anthropologist Tim Ingold was likely read by French philosophers for the first time in the writings of another anthropologist – Brazilian this time – Eduardo Viveiros de Castro, and more precisely in his major work *Cannibal Metaphysics*, published in 2009 in the same collection as the book by Emmanuel Grimaud that we are publishing now.[1]

Since then, much ink has flowed under the bridges connecting the two disciplines. In addition to Tim Ingold and Eduardo Viveiros de Castro, we can mention Marilyn Strathern, Anna Tsing and more generally everything that has been called the 'ontological turn'[2] in anthropology – not forgetting two major national figures who have finally achieved the recognition in their country that they deserved: Bruno Latour and Philippe Descola.

The work by Emmanuel Grimaud that we are publishing now continues this line of conversation, while veering off into strange, new directions – owing both to its method and to its subject. As for the subject, it touches unmistakably on philosophy, and even on its most characteristic branch: metaphysics. After all, it concerns god. What could be more incontestably metaphysical than the question of god? We recall that Kant considered it – along with the immortality of the soul and freedom – as one of the three principal objects of what he called 'dialectic': that is, the intellectual exercise that concerns Ideas, those objects of thought which can never become objects of experience and which are condemned to remain speculative. For Kant, this means they are sources of insoluble controversies (at least as long as they remain theoretical and do not shift to the practical, moral realm).[3]

Yet the very method by which Emmanuel Grimaud approaches this eminently metaphysical object gives the foreword writer pause – just as it will likely trip up some readers: is it accurate to say this is about god? Is it not rather about the human, or perhaps even the machine – or at least their relationships? The reader will see that the book does not resolve this question. Not because it hesitates, but because it does not start with a well-defined idea of what ought to be discussed – whether it is divinity as a reality in itself, religion as a human practice, the social roles required for life in community, or anything else one might suppose in order to explain and situate the phenomena the book reports. It does not carve out in advance a domain of the observable upon which the ethnographer's gaze would rest. On the contrary, instead of assuming a known entity – 'god', in this case – it posits a cobbled-together object that opens an unprecedented space of play and variation: a space where the heterogeneity of beings, prior to their partition into domains, can become the object of exploration. In short, it invites us to experiment with god – or, in every sense of the phrase, in the place of god.

Emmanuel Grimaud's approach is so unusual – both in its form and its substance – that readers curious about metaphysics, who may already find it difficult to appreciate the importance of the author's gesture for anthropology, will find it even harder to perceive its significance for metaphysics. This detailed foreword aims to highlight those stakes. A reader indifferent to metaphysical questions, or conversely determined to discover them on their own, can skip it entirely and attempt to undergo this journey unguided – only to return later to deepen their understanding of the landscape they will have traversed.

***

One of the most striking interests of this book is precisely this: introducing the notion of *experimentation* into anthropology – and perhaps, more broadly, into the social sciences and humanities. Can the question of the essence and existence of god be treated experimentally? Perhaps that, ultimately, is what anthropology can offer philosophy: an experimental treatment of its own questions, such that putting people into its preferred objects (as Ingold said: 'the people in') would be only a secondary effect of a primary operation – replacing essences with assemblages. Here we find long conversations between 'people' that allow us, so to speak, to witness live the adjustments through which a god, a human, a machine, a work of art or a spirit gradually stabilizes.

The experience recounted in this book primarily produces conversations, and conversely, these conversations serve as experimental results. What is anthropology, after all, if not first and foremost a knowledge of, in and through difference – one that proceeds by varying what is considered invariant.[4] And what does it mean 'to experiment' if not precisely to traverse a space of variation in order to reveal thresholds, dimensions and factors?[5] This requires setting up an artificial apparatus that allows parameters to be varied – if only to isolate them, to observe which properties persist when one slider is pushed further, when one element is removed, or when another factor is adjusted.

This is exactly the kind of apparatus created by Emmanuel Grimaud and his collaborator, the artist Zaven Paré (with whom he had already conducted a series of experiments described in *The Day the Robots Will Eat Apples*, 2012). The book we present here is the account of that experiment, delivered as raw as possible so that each of us may enter into the experience – may become immersed in its shifting sands, feel a god arise and vanish here, a prophet there, flashes of divinity elsewhere ... Grimaud and Paré introduced into the world an artefact (a new kind of idol) that applies a sharp twist to certain existing practices (those known as devotional), enough to set in motion what previously appeared frozen, and to reveal certain underlying properties of the object they targeted. This object turns out to be neither god nor religion per se but, rather, a compound of human, artefact and the supernatural.

What exactly are we dealing with? It is a kind of robot supposed to 'represent' (though that word already assumes too much) one of the most important deities in the Indian pantheon: Ganesh, the elephant-headed god. This extremely popular god holds many roles in the daily lives of Hindus – and more broadly, Indians. He is often found hanging from taxi rearview mirrors, as he is the god of exchanges and movement. For this reason, he is also the god of merchants – much like Hermes – and of schools and intelligence.

In India, gods incarnate through all sorts of media: figurines, paintings, giant festival effigies, opulent sculptures in temples, special effects in films, and even automata.[6] The artefact invented by Zaven Paré and Emmanuel Grimaud is, at first glance, just one more realization of a conventional form of the god – understood as contingent relative to his invisible, uncreated essence. Indian polytheism tolerates (perhaps even more than other polytheisms) great latitude in the realization of gods, all in the name of a monotheism sometimes even more abstract than those we think we know. Thus, the creators of this robot could claim a certain tradition. Yet they still committed a major transgression – not from the standpoint of religion, but of science: instead of limiting themselves to studying representations of a deity, the anthropologist dares to fabricate one. And not just any: one strange enough to open a field of hesitation, of inquiry, where one can hope to catch in the act the forces that make divinity – and those that emerge from it. It is not a question of observing or participating, but of intervening by voluntarily disturbing the environment through the sudden intrusion of an artefact – and then recording the effects of this intervention.[7]

The word 'robot' likely evokes a highly sophisticated technological entity. But *Bappa* – as his creators named him[8] – is rather low-tech. As seen in the photograph in Figure 0.1, he resembles a kind of trolley, akin to what's called a hand truck in warehouse handling. Mounted one-third of the way up are symmetrical speakers; at the top, an oval screen flanked by two black ears shaped like elephant ears; and at the centre, a metallic trunk articulated with small discs, giving the impression of a crude tube stripped of its casing, exposing only a ribbed skeleton. Overall, Bappa looks less like one of those eerily humanoid robots promised by roboticists, and more like a somewhat grotesque contraption that makes no effort to hide its mechanical nature: the cables are visible, as are the bolts and even the mismatched origins of its components. It is made from repurposed parts rather than integrated circuits. And yet, when a face suddenly appears on the oval screen, this odd device takes on distinctly anthropomorphic traits: the upright chassis evokes a human stance, the trunk stands in for a mouth, the bars resemble stiff arms pressed to a body without flesh and the wheels act like legs. But these traits are never detached from their zoomorphic and technomorphic counterparts. Curiously, the human face's appearance enhances – not obscures – these other traits. What we face, then, is a cheerful hybrid: an innocuous being that inspires not solemn reverence, but a kind of tender confidence.

This makeshift figure, which looks as though it came from the garage of the postman Cheval, may still rightfully be called a 'robot' – for two reasons. First, it is animated: the ears and trunk move in a rhythm that feels alive precisely because it is irregular. Yet it is not an automaton in the same sense as a clock, because its movements are governed by sensory input. It thus belongs to the family of cybernetic machines. Second, this object is tele-operated – that is, remotely controlled. Like the rovers sent to collect minerals on the Moon or the unmanned submarines exploring the ocean floor, Bappa is operated from a distance. However, the operator does not control the robot's movements. Instead, they lend it their face and voice – what one might call his conversational *persona*.

The operator is an individual who agrees to embody the god, to speak as the god, within the context in which the device is placed. Their face appears on the screen; their voice emerges from the speakers – and a free interaction then unfolds with any willing participant. At the start of the experiment, the operator was hidden from those interacting with the machine – placed in another room, or even a different location, so the source of the manifestation was unknown. But the experimenters soon realized this created a paradox: the concealment made the god-role even more suspicious. It complicated the game of incarnation and made the desire to speak in the name of the deity seem less credible. Eventually, they installed the operators in a separate but accessible space, open to anyone curious enough to look. This move sidestepped the question of trickery or artificiality.

Far from undermining the machine's function as an idol of Ganesh, this transparency actually deepened the experience – like a polished surface reflecting back the question, 'Where is God?' The fact that the incarnant was not God seemed metaphorically acknowledged by the fact that the robot was not the incarnant. But these dual acts of delegation allowed for the possibility that God himself might tele-operate the incarnant, just as the incarnant tele-operated the machine. And indeed – that's exactly what happened. God returned to the machine, as if by surprise.

One cannot help but be struck by how, in this seemingly modest set-up, the divine manifests – here or there. Some of the 'incarnants' suddenly find themselves swept away (the most spectacular case being the young Varsha, described in the chapter 'Ganesh Cacophony') and are surprised by what they say. They express insights or wisdom they didn't know they had, as if inspired by a voice that might very well be that of the god.[9] More generally, it is surprising to see how, once the idea of pure trickery is dismissed – once interlocutors of Bappa understand that this isn't about deceiving them, but simply about attempting something authentic, ingenuous – they genuinely begin to address the god. They grant the entity an existence that cannot be reduced to mere representation.

This sometimes leads to remarkable paradoxes. For instance, when someone questions the god's existence – asking more precisely, 'Where are you?' – the incarnant responds, 'I am inside you, I am in your heart. Do I not exist through your devotion?' This answer satisfies the devotee. After all, as we've already noted, the god resides in a plaster idol thanks to the same cause that allows the incarnant to be simultaneously inside the machine and in the role of the god – namely, the goodwill of the devotee, or more generally, a shared agreement among the participants.

But what a curious mechanism! Here is a being accepted as real only by insisting it is a representation. A similar paradox arises with a woman who uses the opportunity to speak to a responsive idol – finally, one that doesn't remain silent – to voice a long-held grievance. She suspects the god of not existing because, despite her faithful offerings, she has not been rewarded. It is as if the 'false' trick of the robot enables her to speak about a deeper deception: that of the god himself. But she can speak of it only through a paradox – by accusing an entity of nonexistence, she causes it to exist, if only to reproach it for lacking being.

This shows that the opposition between 'thing in itself' and 'object of representation' is ineffective when trying to grasp what occurs in the emergence of the divine. The god is neither a self-sufficient reality that exists independently of any set-up, nor merely the intentional object of internal representation or collective practice. Rather, god is an effect – as in Doppler effect, Joule effect or Faraday effect. Not an illusion or mere appearance, but a fully real, observable phenomenon – provided certain conditions are met within an experimental framework, or at least one that attempts to capture such conditions.

God does not *exist* – God *happens*.

God happens when a course of action deviates, when something believed to be carried by inertia is suddenly struck by something outside itself – when something is inexplicably heightened. God arrives when there is *alteration*. This leads to a strong metaphysical thesis – one that Emmanuel Grimaud never states explicitly, but which clearly emerges from the material he presents: *Otherness precedes Being*. An entity exists to the extent that it alters the course of another (which, in turn, is revealed in its difference at that moment). This thesis closely resembles what Bruno Latour developed (speaking explicitly of 'the other as other') with similarly subversive goals: detecting entities that are constantly at risk of being misrecognized (a legal injury mistaken for psychological trauma, for instance), and rethinking beings as entities that navigate an ontological ambiguity. As if every being ultimately tears itself from a murky background.[10] Here too, it becomes clear that divine and human, technological and artistic, electric and mechanical – are not separate domains. They are not to be kept apart or reduced one to the other (e.g., god to human, or to artefact), but are instead effects, events, phenomena – equal in the sense that they all emerge from a common condition: the need to constantly rise up out of a troubled, impure, radically essence-less background – a background that is neither god, nor man, nor machine, nor any fixed category.

As with Latour, entities in Grimaud's account exist only on the condition that they fulfil a specific set of requirements. Anthropology's task, then, is to describe for each context what these requirements are. In the case of Ganesh, this 'specifications sheet' emerges in a particularly interesting form: as we will see, an incarnant is all the more effective the more they invent (or receive) new roles and functions for their god. Thus, performing the divinity – much like Judith Butler's concept of gender as performance – means not merely reproducing or imitating an established figure but, rather, expanding it, directing the god towards new incarnations. This may, in fact, be a general principle: a being truly exists only if it discovers more of itself than it previously knew. A being that merely copies itself – one that places too much faith in a Platonic ontology – eventually withers into a caricature, a pale imitation of itself.

***

Grimaud's experimental anthropology thus marks a departure from what has been called the 'ontological turn' in anthropology (e.g., Viveiros de Castro, Descola). Anthropologists in that tradition aimed not to deny the existence of the entities acknowledged by the people they studied – spirits, gods, forces etc. – but to avoid reducing them to mere mental representations, as if opposed to a single, true nature revealed only by Western science.

But Grimaud's approach assumes that point is settled. The problem is no longer whether these beings are real but, rather, how not to accept too quickly the hardened forms we call gods, humans, machines, plants, viruses or words. His apparatus invites us, so to speak, to descend one level lower – to dwell within a magma of still-unstable entities, entities that are overdetermined and underdetermined at once, or that hover at what Grimaud calls their 'zero-point'. Thus, when divinity emerges in certain incarnations of Bappa, it cannot be attributed simply to the god – because we know the incarnant is a fellow human. But neither can it be attributed solely to the human, because they feel themselves surpassing their usual capabilities. Nor can it be explained by the device's power of suggestion, which alone is not sufficient to produce the effect – it requires the intervention of both the human and the godlike.

What we are facing, then, is not a hybrid being, but an equivocal one: multiple by overdetermination, because several identities converge on it, and also multiple by underdetermination, because none of those identities is fully formed or autonomous. We are thus in a space where being and making-be collapse into each other. The question is no longer whether we should be liberal or strict in our ontological judgements. It becomes: how, and for how long, can we remain within this molten ground and observe the birth of entities? How can we track the conditions, gradients and parameters through which they harden, stabilize – or destabilize – or stall altogether, reaching what Grimaud calls a zero-point? From there, he reinterprets both metaphysics and experimentation. This experimental turn in anthropology is not a regression – it is a deepening and radicalization of the ontological turn. And it is for this reason – among others – that it becomes of interest to philosophy.

***

We already see the strong metaphysical lessons Grimaud's apparatus brings to light: beings are indifferent to the dilemma of representation versus reality (they may assert themselves entirely outside of that dichotomy); otherness precedes and enables being; entities are not confirmed substances, but performed effects; they arise out of a murky, indeterminate background, always emerging relationally and differentially. But there is yet another philosophical teaching to draw from this astonishing experiment: the intimate connection between being and force – or, rather, forces in the plural.

One of the most spectacular effects of Grimaud's set-up is that, in place of what previously froze a practice, it inserts something that reopens it, makes it negotiable again, modifiable, malleable. This is particularly evident in one of the early interactions involving the makeshift idol, between the fisherman's wife and the rigid Karan acting as the incarnant (see the chapter 'Divine (Im)posture'). This is the same woman we discussed earlier – who initially challenges the incarnant's authenticity and questions his legitimacy to present himself as the god. But eventually, her reproach shifts from the operator to the god himself, as if she had grievances to settle – grievances she could articulate only because she was speaking to someone pretending to speak in the god's name.

The apparatus literally enables negotiation with god about god – an opportunity to see under what conditions one accepts the deity, in exchange for what, under which terms – literally: in which words, what formula of peace. By inserting this strange artefact in place of the idol, the anthropologist opens a space of play, of negotiation, of forces yet to be resolved, where once there appeared to be settled questions. I say 'appeared', because it truly is only an appearance. The relationship to god, the identity of god, of the human and of the other entities that populate our world – nations, drives, media, whatever else – is never definitively decided. It is always being negotiated, conditioned, put to the test. And yet, in the case of divinity at least, these operations generally unfold outside our interactions with the material mediators of the divine. Churches, for instance, rarely allow for the public, material traversal of theological doubt. They are themselves the result of theological disputes, yes – but they present only the stabilized, resolved, dogmatic versions. Bappa, by contrast, is like a synchronization table in music: you can vary this parameter, turn that knob, shift across the scale of possible states of one's relationship to god – and of god's very existence.

Above all, Bappa allows us to reach what Grimaud calls the 'zero-point.' This is a fundamental concept. It designates moments of blockage, where the operations of negotiation, stabilization and requalification break down – as in a computer 'bug'. The conversation stalls on a kind of indeterminate plateau where nothing seems to happen anymore. The most striking example is that of Karan himself, who, faced with the woman confronting him, is reduced to repeating the mantra 'Om', unsure how to proceed. Such points are essential because, Grimaud argues, they are where invention begins. You must push a practice to its breaking point – lead a routine to where it becomes unworkable – to find yourself forced to reinvent what is a god, what is a human, a machine, a work of art, a science. But who reinvents? Who else than human beings, the social actors?

One of the most fascinating teachings of this book is that the question of who invents cannot be answered in advance. When the incarnants and their interlocutors hit a wall – when they can no longer perform their roles – is it the human who invents a better performance? Or the god who finds a new way to be incarnated? Or the anthropologists, who invent an improbable observatory? It's impossible to say ahead of time. And that's not the point. The point is that invention becomes necessary. It's that – or nothing. The subject of invention comes after its initiation.

What is invented after the zero-point, Grimaud calls a 'boundary-plane of experience'. He defines this concept as: 'Any configuration that brings an existing being to the limits of its perception or its capacity to act – to its blind spots.' He gives vivid examples: 'Isn't a mountaineer immediately confronted with a boundary-plane when the ground shifts to a vertical plane, forcing the use of ropes, gear and new techniques? Isn't the water's surface, for a fisherman, a boundary-plane in the sense that he must constantly interpret signals and invent traps to catch his prey?' Anthropology, Grimaud suggests, consists in settling upon such boundary-planes of experience – or even in creating them. But here we must pause on one particular remark: to define anthropology as the exploration of 'boundary-planes of experience' casts new light on its unique intimacy with metaphysics.

Typically, anthropologists have assumed that they encounter metaphysics in their subjects in the form of systems or meaningful structures – whose content overlaps with topics traditionally assigned to metaphysics. Even the celebrated – and often turbulent – wedding of philosophy and anthropology, which the 'ontological turn' recently celebrated, remains grounded in this approach. From Marcel Griaule's dialogue with the Dogon sage Ogotemmêli to Viveiros de Castro's 'Amerindian perspectivism', anthropologists have been reconstructing alternative metaphysics – in their concepts, categories and forms – to those that dominate Western philosophical history.

Grimaud, while aligned in some respects with this trajectory, introduces a radical bifurcation. In his framework, metaphysics is no longer defined by its concepts, topics or doctrinal structures. Instead, it consists in unstable trajectories of interaction across shifting states where beings may stabilize temporarily, but only after having navigated impasses – zero-points – where previous modes of being collapse. And this is a fully legitimate definition of metaphysics. Because it is not absurd to define metaphysics as what one does once all routines have broken down – where one encounters the famous Aristotelian 'thaumazein', the astonishment triggered not by the strange, but by what normally fails to astonish us. We enter metaphysics when we traverse a zone in which all known frameworks are disoriented, and yet we must continue forward.[11] Thus: anthropology is experimental because it confronts these boundary-planes – and it touches metaphysics precisely in doing so.

If the experience recounted here belongs to metaphysics, it is not because it discusses a transcendent object called god, but because it leads god to his zero-point. Not every relationship to god, not every discourse about god, is necessarily metaphysical. A relationship to god becomes metaphysical only if it passes through such a point. The divine is not, by nature, an object of metaphysics – no more than anything else is. But when we reach the zero-point – or better, when we build a device that allows us to reach it, or even better still, when we reach it in multiple ways and map out the different zero-points and the zones that lie between them – then we are truly engaging in a metaphysical experiment. Metaphysics is that which allows us to carry on after encountering the zero-point. It is what allows us to reconstruct a viable plane of functioning when all previous operations have collapsed.

Grimaud introduces a distinction between ontology and metaphysics that should interest contemporary philosophy, especially thinkers like Quentin Meillassoux and Tristan Garcia. He does not define metaphysics as an ontological commitment, or as meditation on being or non-being. Instead, he defines it as a point of ontological neutralization, a moment when the question of what beings are and whether they even exist is suspended, because the question of their practicability arises, of the possibility of continuing interactions, of constructing a space where a course of action can be pursued beyond its blockage. In short, if Bappa is a metaphysical contraption, it's because he is an experimental artefact of a certain kind – a machine for confronting the zero-point.

***

It is as though the robot has absorbed, within the idol itself, all the tensions, power dynamics and complex negotiations that normally surround an idol – giving it its function and ontological status. Plaster idols are, in fact, entangled in many shifting power struggles: attempts to claim god for political, cultural, social or personal goals; doubts about the god's effectiveness, value or existence; debates about the best uses of religion; even theological disputes. All these contribute to what idols are and how they take shape – think, for instance, of the Baroque style and its debt to the Counter-Reformation. Yet, such dynamics usually unfold outside the idol. Idols do not reflect upon the conditions of their own existence.

However, if the idol is indeed an idol, if it functions effectively as such, it is because at least at one point these controversies have stabilized and we do as if the fight could be suspended. In contrast, Bappa brings all these controversies into the idol itself. In this way, he replaces what might be called a *resolution-oriented* artefact with a *problem-oriented* artefact. He internalizes all of these controversies. He does not work like a traditional religious artefact, which can function only once certain problems have been resolved. Bappa works precisely by generating a space where these problems can be played out. This is undoubtedly due to its experimental nature: not a fixed embodiment of god, but a field of variation for exploring the limits of divine experience. To experiment is also this: to internalize external forces in a bounded space where variation may not always be controlled, but can at least be marked, tracked and mapped.

This proposition leads to two philosophical consequences, which I would like to emphasize as we conclude – one concerning the philosophy of technology, the other, metaphysics. Both converge on a single word: *composition*.

From the technological point of view, we note that Grimaud proposes a view of technical objects not as solutions to human problems, but as occasions to invent new problems. His 'god-making machine' is not meant to replace traditional idols. Rather, it aims to turn the entire ensemble that allows idols to exist into a terrain for exploration – a space where worshippers, intermediaries and even perhaps the god himself can rediscover the terms of their existence. This kind of object resembles what has been called *speculative design* in the design world, or speculative architecture in the field of urban planning.[12] In design, this refers to objects not aimed at seamless functionality, but that challenge the user to question their needs. In architecture, it refers to structures less focused on providing shelter and more on provoking reflection on the act of dwelling itself. With Bappa, we encounter a new category: speculative robotics – not in the sense that it speculates about god, but in the sense that it generates thought through its technical presence.

Although Grimaud does not position himself explicitly within this lineage, his work aligns with a now well-recognized current in the philosophy of technology, which opposes the paradigm of simulation with the paradigm of composition. In the simulation model, technology seeks to replicate human capacities – to perform actions so that humans no longer have to. In the compositional model, technical objects become collaborators, enabling new possibilities for life, action and thought. A striking predecessor of this approach is the Tamagotchi – the handheld digital pet shaped like an egg. Its screen became, in a sense, the Levinasian face of a virtual creature, one the user had to care for – or else let it 'die', i.e., become unusable. This object did not simplify human life; it complicated it. Unlike a dishwasher that relieves us of a chore, the Tamagotchi created one (feeding, caring, attending).[13] Similarly, Bappa does not aim to 'do god' better than plaster idols. Rather, it inserts a different kind of game into the religious structure – a new but related form of engagement. Thus, the purpose of technology, in this view, is not to reproduce beings without us, but to expand the list of beings so that we might discover new relationships, new companionships and even new dimensions of our own being.

This departure from the logic of simulation should alert philosophers. For decades, philosophers have turned to machines for insights into age-old questions. The most emblematic case is the mind. Many thinkers believed that computers would help us revisit the question: 'What is thought?'

Alan Turing famously proposed a test to determine if machines could think – not by probing internal states, but by their ability to simulate conversation convincingly. His idea was: if a machine can imitate human discourse so successfully that a human cannot distinguish it from a real person, then it 'thinks'. Thus, the question 'Do machines think?' becomes, in effect, 'Can machines simulate us well enough to fool us?'

This simulation paradigm became dominant in how philosophy has related to machines. Had Grimaud followed that model, his experiment might have aimed to build a robot capable of convincingly simulating god. In fact, Descartes himself once imagined using science to manufacture miracles on command, thereby shoring up faith through artificial signs. But Grimaud and his collaborators took another route. They built a device that allows humans to explore what it costs to simulate a god. It's up to them to make the god. Like a mask, a chasuble or a mitre, this artefact functions as a divine prosthesis, enabling a human being to acquire more-than-human qualities, provided that he is capable of performing the god in return. The question is not whether god can be artificially reproduced, but what combinations of the human, the divine and the mechanical can work. It's not a question of reproducing a phenomenon, but of opening up new possibilities for action. In short, it's not about a relationship of *substitution* of one being (the machine) for another (the divine), but a relationship of *composition*.

***

The concept of composition has a second philosophical implication – this time aimed squarely at metaphysics. What Bappa allows us to see is this: within a single object, we can gather and expose all the external forces that normally surround an idol, giving it its role, its identity and its ontological status in a practice. We begin to see, rising to the surface, all the competing forces that vie for the god – each trying to enlist the deity for their own cause. For example, in the chapter 'Divine (In)Justice', a participant named Nitin tries – against the will of the current incarnant – to recruit the god for his anti-corruption campaign. Others attempt to align the deity with LGBTQ+ rights, still others with ecological causes. Every time, these uses encounter resistance from other forces in the scene.

And yet – it would be a mistake to think that the god is just a passive object, a pretext or an empty vessel for these agendas. As Grimaud rightly emphasizes, if people struggle to capture the god, it's because the god contains a force – or multiple forces, sometimes a force of detachment, helping believers distance themselves from worldly troubles; sometimes a force of collective mobilization, rallying people into shared action; sometimes a force of destruction, legitimizing violence in the name of justice; or a force of protection, directing attention and care to vulnerable beings – like plants, animals or the Earth itself. Here, 'force' means: that which makes a certain kind of action possible – that which acts on other forces in a particular way. Mobilizing forces bring together; Destructive forces separate or negate; Detachment forces abstract and isolate; Protective forces guard or stabilize. Bappa, then, is not a fixed form but an apparatus that reveals these interplays of force – a dynamic field where forces express themselves, clash, intertwine.

And then something essential – something even extraordinary – happens. As the fixed relationships between established entities dissolve in this shifting field of force, the very categories we thought we were coordinating – 'human', 'god', 'machine' – begin to lose their secure boundaries. We can no longer assign specific forces to specific identities. The god, the human and the machine no longer own their roles. Instead, their very emergence depends on the configuration of forces. It is not that a human wields divine power, or that a machine pretends to be alive. Rather, these forms are the result of certain force compositions. Where one type of force dominates, 'a god' may appear. Where another constellation arises, 'a human' emerges.

And still another may give us 'a machine'. Forms come later. Forces come first.

The traditional categories are not prior to the experience – they are outcomes of the experiment. And here Grimaud's work becomes invaluable: it not only presents a theoretical truth, but it allows us to observe the spectrum of forms that emerge from force-composition – in real time, within a single experiment. This is something history can teach us in retrospect. But Grimaud's set-up lets us experience it directly, within a single experimental frame.

It is precisely at this point that Emmanuel Grimaud invokes Nietzsche and Foucault, both filtered through Deleuze's reading. Nietzsche's *Übermensch* (Overman) and Foucault's 'death of man' both suggest that 'the human' is not to be defined by biology, morality or existential condition alone – but as a historical configuration of forces, one that may disappear. This does not mean that human beings will vanish as a species, or that empathy, duty and inner life will cease. It means that what we currently call 'the human' may cease to exist in its present form, because the forces that sustain that form may realign.

Grimaud quotes from Deleuze's famous essay on Foucault, which opens with this key principle:

> Foucault's general principle is: every form is a composition of forces. Given a set of forces, we ask: with what external forces do they enter into relation and what form results. There are forces in man: to imagine, remember, conceive, will ... But such forces do not presuppose 'man' as a form. They only require sites, points of application,

> regions of existence. Just as forces in animals (mobility, irritability) presuppose no particular form. The key question is: what other forces do these forces enter into relation with. From that interplay, a form emerges. Even over short periods, the human has not always existed – and will not always exist. For the form 'Man' to appear, forces in man must connect with very particular external forces.[14]

Grimaud sees Bappa as a kind of observatory for watching, in real time, the composition of such forms. His apparatus lets us think differently about what is happening to humanity in the age of proliferating machines. And this leads to what may be Grimaud's boldest and most forward-looking claim: the new technologies do not threaten to replace humanity.

They demand that humanity recompose itself. The question is not whether humans will survive or be outpaced by machines. The real question is: what forces are being released now that old forms are destabilized, what alliances can be made between human, divine and machinic forces; what new configurations will give fresh meanings to words like 'god', 'machine' and 'human'.

This shows just how much this anthropology remains an anthropology – that is, despite everything, a discipline concerned with the question of 'the human' not in the sense of an essence but, rather, in the sense of an equivocal figure that goes as far as the notion of the species goes. Lévi-Strauss once said that anthropology's task has always been 'to dissolve Man', not to affirm him as an object of knowledge. Grimaud continues this legacy. On the other hand, this discipline has the rare ability to help us understand how, under what conditions, in what circumstances, this form that is the Human appears as an effect, an effect that is negotiable, revisable, equivocal.

But we can't grasp the precariousness of the human, its contingency, its negotiability – a profound truth that opens up a veritable field of fertile empirical study – without at the same time freeing all beings from the ontological straitjacket in which the human sciences have defined themselves – without, in short, going through an authentic metaphysics, i.e., giving definitive leave to all the ontological securities which distribute true beings on the one hand, and mere appearances on the other. That is, you must give up the belief that reality is neatly divided into 'real' things and 'mere' representations. There are only forces and effects. And that is the experience this book offers us: an edge-of-reality experiment. A limit-experience that is possible only when anthropologists of a new kind tinker – carefully, ingenuously, obsessively – with a little apparatus that brings each formed being back to its zero-point – even god.

Notes

1. Eduardo Viveiros de Castro, *Métaphysiques cannibales* (Paris: Presses universitaires de France, 'MétaphysiqueS', 2009).
2. See Martin Holbraad and Morten Axel Pedersen. *The Ontological Turn: An Anthropological Exposition* (Cambridge: Cambridge University Press, 2017).
3. Immanuel Kant, *Critique de la raison pure* [1781], Deuxième partie, deuxième division, 'Dialectique transcendantale'.
4. On this point I would refer you to a text in which I attempt to develop philosophically a radical epistemology of anthropology in line with the teaching of Eduardo Viveiros de Castro: P. Maniglier, 'La Vérité des autres', in E. Alloa and E. During (eds), *Choses en soi: métaphysiques du réalisme* (Paris: Presses universitaires de France, 'MétaphysiqueS', 2018).
5. Claude Lévi-Strauss argued that cultural variation made it possible to introduce into the human sciences an experimental dimension that the nature of the subject of these sciences prevented them from introducing, if only for ethical reasons (see in particular *Anthropologie structurale* (Paris: Plon, 1958 [reissued by Pocket]), 333.
6. In a previous book, Emmanuel Grimaud worked on religious animatronics and the use of machines, special effects and automata on the ritual platforms in India. See *Dieux et robots: Les Théâtres d'automates divins de Bombay* (Apt: L'Archange Minotaure, 'L'Âme du monde', 2008).
7. The word 'intervention' here evokes an eponymous poem by Henri Michaux, the opening line of which I quote: 'In the past, I had too much respect for nature. I stood in front of things and landscapes and let them be. No more, now I will intervene. So I was in Honfleur and I was bored. So I decided to put a camel on it. H. Michaux, 'Intervention', in *La Nuit remue*, in *Œuvres* (Paris: Gallimard, 'Bibliothèque de la Pléiade', vol. 2, 488) could be a manifesto for the experimental method in the humanities.
8. Bappa means 'father' and is a name often given to Ganesh in the form of Ganapati Bappa.
9. But the originality and the interest of the device, as we shall come back to it, is that it does not require this voice to be that of the god, because the incarnants as well as their interlocutors can constantly attribute it to something else, to the talent of the incarnants, to the ability of the consultants to steer the speech in the right direction, or even to the effectiveness of the device itself, a bit like actors entering their role when they put on their costume or any individual starting to speak like a king as soon as he wears the crown. It even seems to me that the voice can all the more be that of a god if it is not condemned to be so and if it retains this ontological freedom, this mutability of qualification: we are all the more willing to recognize a divine effect in it if we are not obliged to do so. This is one of the most interesting lessons to

be learned from this experience: that beings are born by first reserving the possibility of not being taken too seriously, by maintaining themselves in a kind of ontological uncertainty guaranteed by the superimposition of qualifications. We'll come back to this later. One of Grimaud's major theses in this book is that the technological hybridization to which we are increasingly condemned will give rise to many new entities, rather than simply replacing one kingdom with another. In short, this emerging god is the harbinger of unknown entities that our technological civilization is preparing in place of the promised great replacement.

10. See B. Latour, *Enquête sur les modes d'existence* (Paris: La Découverte, 2012); on 'being as other' in particular, 172.
11. This experience is reminiscent of what Tarkovsky stages under the generic name of 'zone' in his film *Stalker*, which would explain why the metaphysical dimension of this work should be sought not only in its contents of meaning, but also, and above all, in the regime of experience it presents (see Jeanne Etelain, 'Qu'appelle-t-on zone? À la recherche d'un concept manqué', *Temps Modernes*, no. 692 (January-March 2017): 113-35). Moreover, I myself have defined metaphysics as that practice which begins at the point where another practice not only encounters failures and obstacles, but becomes impracticable because of its very effectiveness. See Patrice Maniglier, 'Manifeste pour un comparatisme supérieur en philosophie', *Temps Modernes*, no. 682 (July 2015).
12. For speculative architecture, read Dominique Rouillard, *Superarchitecture. Le futur de l'architecture 1950–1970* (Paris: Éditions de la Villette, 2004).
13. On this use of artificial intelligence, let me refer you to the last section of a text I wrote: 'Calcul à cultures: le "structuralisme" dans l'histoire de l'intelligence artificielle', *Education et didactique*, no. 1(3), Rennes, Presses Universitaires de Rennes (November/December 2007): 81–107.
14. G. Deleuze, *Foucault* (Paris: Minuit, 1986), 131 ('Annexe: sur la mort de l'homme et le surhomme'/'Appendix: On the Death of Man and Superman').

PREAMBLE

A God in the Machine?

As the celebrations for the Ganesh festival (*Gaṇeśa Chaturthī*) approached, it was difficult to find a *paṇḍit* (officiant) who was available. Shailesh, the production assistant, took it upon himself, however, to scour Mumbai to find one. It was out of the question, in his view, to start the experiment without an inaugural ritual involving a Ganesh idol. Ours wasn't exactly ordinary, and we didn't know if the *paṇḍit*, under such conditions, would bless the project like any other undertaking. Upon arriving at the house in the heart of Mumbai, which we had rented for the duration of the shoot, he asked where our idol was. Karan, who was willing to be its first incarnant, pointed at the machine.[1] Nobody had told the *paṇḍit* that it was a robot. Equipped with a thermoformed mask and a trunk, no doubt a bit abstract, it was elephantine all the same. There was no mistaking it, the machine had something a bit Ganesh about it.

It made no difference to the *paṇḍit* that the idol was a machine. True, he was paid as for any other ritual, but that was no reason not to wonder about what was at stake in the endeavour. Rituals to open the eyes (*prāṅa pratiṣṭha*) are performed on idols so as to breathe life into them. Inauguration rituals (*muhūrta*), for their part, are designed to attract favourable forces at the beginning of any undertaking. Nothing prevents the combination of the one with the other. The *paṇḍit* sat down and unpacked all his paraphernalia. A *jajman* (sponsor of the ritual) was needed. The designer of the machine, Zaven Paré, was thus appointed. His long beard, which made him look like Christ when in Paris, made him look like Brahma, the creator, once in Mumbai. He was told what to do. Each gesture, each operation, was explained in a very didactic way by the officiant. We had only to obey. And at the moment of the final blessing, all operations were oriented towards the machine. A garland was placed on it, and on its forehead the officiant invited us to apply powder using the thumb at the locus of the third eye (*ājñā*), in order to open its 'subtle channel.'

The discussion that followed the ritual began from a question posed to the *paṇḍit* about freedom in making idols. To what extent do we have the right to invent a Ganesha or to give him a different form? And in what category of idol should we put the machine?

'Self-generated (*svayambhū*) is he who comes and goes of his own will. Not *svayambhū* is he who is born in his mother's uterus; not *svayambhū* is he who is afraid of death. *Svayambhū* is he who is neither born nor dies,' the *paṇḍit* replied. 'He comes when you need him, and leaves when your need vanishes. But when you invoke him, he comes ...'

'Are you saying that this machine could be considered a self-generated form?' I asked him.

'A self-generated form is not subject to the laws of birth and death. Its power is thus unlimited,' the *paṇḍit* replied.

'Doesn't the god Hanuman also have unlimited power?' Karan added.

'Hanuman was born from the belly of Anjani, his mother,' the *paṇḍit* said. 'He is not *svayambhū*. The Ganesha that you are going to worship on the 29th September [the date of the first ritual of the *Gaṇeśa Chaturthī*] is the son of the god Shiva and the goddess Parvati. That Ganesh is not self-generated. He was made from the sweat of Parvati. But the primordial Ganesha is he who came when this universe was created by the gods Brahma, Vishnou and Mahesh. It is the ultimate Ganesh, the one that we just invoked by the ritual that I have performed. The Ganesha that you will venerate in four days' time is a little incarnation of Ganesha himself. In fact, it is one of his many incarnations.'

'Do you mean that it is sacred (*āradhya*)?' Karan asked.

'No', the *paṇḍit* answered patiently, '*āradhya* means the one without whom nothing can be done, the only one you worship. The Ganesha who is the son of Shiva and Parvati is different to the Ganesha who existed even before the Universe had been created, who will still be there once the Universe reaches its end. The Ganesha who is the son of Shiva and Parvati is different, he is the one you will worship on the 29th before immersing him,' he said, pointing towards the robot. 'As for the Ganesha whom we just venerated, for him there is neither birth nor death.'

Up to this point, none of us had asked ourselves how to bring a machine into the pantheon of existing divinities. Karan, who was preparing himself to be the first incarnant, seemed to have understood what we could draw out of the *paṇḍit*'s explanation, as well as his gesture towards the machine, which had sown confusion: being a machine, it aspired to perpetual movement, thus to the state of *svayambhū*. Shailesh, the production assistant, however, disagreed.

Karan turned towards me: 'Did you understand it? That which is neither born nor dies, is called *svayambhū* (self-generated.) We don't have to immerse our Ganesha in the sea because it is *svayambhū*.'

'Not at all!' Shailesh exclaimed.

'He was made,' said the *paṇḍit*.

'Is it not the goddess Mataji who created it?' asked Karan.

'Even if it had been created by Mataji', the *paṇḍit* explained, 'it will be destroyed. It is natural. The offering was made [to the robot].'

'Are you the one who gave birth to this?' Karan asked Zaven, the designer of the robot, pointing at it. Zaven nodded.

'All that is created by man, must be destroyed by man!' the *paṇḍit* concluded.

The indetermination as to the ontological status of the machine could have passed in silence. The ritual had taken place in a seamless sequence of transactions, gestures, ritual concoction, offerings, transfers of powders and substances, diluting the question of the status of this particular medium with regards to others. But if the *paṇḍit* gave the impression of having accepted it as it was, regardless of how ambiguous it was, he had not settled the question, and had simply given us elements to decide for ourselves. A god in the machine? The question was, rather: which god in which machine, and for what purpose?

To consider the machine as a self-generated form, without a creator (*svayambhū*), would have been a gross theological hypocrisy, since it had been designed by humans. Attributing its birth to Mataji (The Goddess), that great matrix at the origin of everything, was probably more acceptable, but could in no way end the debate. We agreed, with the *paṇḍit*'s blessing, to proceed with our endeavour, which meant that our invention had to follow the same path as any idol, destined to be immersed. Never could we have imagined having to evoke this outcome from the very start. In spite of its vague resemblance to a well-known god, up until then the machine had been floating in a categorical blur. The invocation ritual had only seemingly put an end to the uncertainty. Theologically speaking, it was now 'an idol like any other', a 'support among supports'. If the officiant had pointed his finger at the machine, it was the abstract form of Ganesh at which he aimed, the eternal form, beyond the machine, which was only a gateway and which was also to suffer the same fate as the classical supports in terracotta or plaster. Everything that is living dies in the end, 'everything that is created by man is destroyed by man'.

By invoking the principle of the unengendered to animate the robot (as with any idol), the *paṇḍit* had enlarged the field of possibilities in order to resolve the problem of its nature in another way, by pushing the form that we presented him with onto the side of temporary forms and acceptable substitutes. Nevertheless, he could clearly see that it was not an idol like any other, since it was endowed with its own animating principle. It was, moreover, a little more experimental than usual, since no Ganesh had ever gone around in this form before. For the *paṇḍit*, the fact that people were having fun experimenting with this apparently serious thing called religion was not a problem. He even helped to define the metaphysical terrain of the experiment, what its field of play would be, for it was not to be formulated in any old way, after more than two thousand years of Brahmanic speculation around the notion of incarnation, as well as that of the substitute. There was perhaps no serious difficulty in inventing an experimental form of god in a universe which has produced a multitude of them, and which has seen, since Vedic India, Brahmins treating the phonemes of the Sanskrit language as gods before resolving to give their deities human forms. But no one could say, before doing the experiment, whether a little more tinkering, a bit of electricity, artificial intelligence and a tele-operated voice, would or would not generate something else, something unknown, something not entirely assimilable to the paths of the sacred already encountered. The *paṇḍit* had understood this, but had not bothered to pay attention to it until Shailesh explained the principle to him: 'The incarnant has their face transmitted to the robot via a webcam and anyone can play the god, during an interaction,' the production assistant went on to say.

'Interesting!' The *paṇḍit's* curiosity had been piqued. 'So it can be a human being as well? But how will the incarnant be able to answer the questions people ask without prior knowledge? People can ask anything. As you know, there are many people who are a bit stupid, but there are also some who are very intelligent.'

'That's easy,' Shailesh replied. 'For example, if someone is unemployed, he will ask the god, "Please help me since you are God, etc." They'll always be general questions.'

'That's right,' the *paṇḍit* affirmed.

'God can provide a simple answer to these kinds of problems. The answer could be, "I only help those who help themselves" or "If you work hard, you will be rewarded." The people incarnating god should express themselves as simply as possible. They should not say anything complicated or negative.'

'And if someone asks you, for example, how many hymns there are in the Atharva-Veda, what will you answer?' the *paṇḍit* asked.

'Google!' Shailesh exclaimed.

'How do you use Google then?'

Shailesh whipped out his phone. 'Mobile ... mobile ... it takes ten seconds Guruji ... this is modernity. For any question, the answers are on Google.'

It probably wouldn't be as simple as Shailesh imagined, but one thing was certain: we didn't want all the incarnants to be *paṇḍits*, even if their knowledge of divine matters would have made them the most likely candidates at first blush to embody God. Among the Indian members of the team, each had their own idea about how it would go, even if no one could predict how the conversations would turn out. The *paṇḍit* was asked if he could return for the immersion ritual. He offered to do the ritual in advance, to save time and not have to come back, as his schedule was very tight.

'Do you have yogurt and sugar? For the immersion, you will take this coconut, it has to be immersed in the sea too,' said the *paṇḍit*. Shailesh nodded. 'Will you be able to do it?' the *paṇḍit* asked. 'You do it only if you can.'

'Yes, I will,' Shailesh said.

'Put this cloth in a bag,' the *paṇḍit* ordered him.

Shailesh returned with a plastic bag.

'No, not in a plastic bag, because we have made the *Abhiṣeka* [idol bath] by pouring the mixture [*Pañchāmṛta*] of the five substances (yoghurt, milk, honey, sugar and butter) on it. You will have to wash the cloth afterwards, not now. Wash it tomorrow and let it dry completely. Keep it like this in this box and the direction should be east-west. Keep it anywhere, but be sure to keep it in that position. You will need to dip it in the water with the machine.'

'With the machine?' gasped Shailesh.

'Yes.'

'Okay.'

'Well, it's done, I've done the immersion ritual.'

'Okay.'

'You understand? Let it be immersed in the sea. This water you can put on plants. Then eat these offerings and the betel nuts, it's a good omen.'

Note

1. Translator's note: Emmanuel Grimaud uses the term '*incarnant*' to mean the one who incarnates the machine, which I have taken the liberty of using directly, considering it preferable to 'the incarnation' or 'the embodiment.'

ACKNOWLEDGEMENTS

The *Ganesh Yourself* experiment, before it became a book, was a collective adventure, involving many people, in India as well as abroad. Bappa would have not taken this shape without the genius of my friend, the artist Zaven Paré, who gave the device its concrete form, and Arnaud Deshayes, who accompanied me all these years and especially during the shooting and editing of the film *Ganesh Yourself* (Arte/Rouge international, 2014). Many thanks to Nadia Turincev and Luciano Rigolini, who made the film possible, Anand Tharani, Shailesh Rai, Praveen, Karan, Jyot, Varsha, Harish, Modi, Nitin Pawar, Anand Pendharkar, Professor Girish Kumar, Suresh and Sushant Sandal and all the people who appear in this book, for their insatiable enthusiasm, curiosity and will to make things happen. I thank Patrice Maniglier, who saw better than I could all the philosophical potential of this experience and encouraged me to publish the French version of the book in his 'PUF MétaphysiqueS' collection, Martin Holbraad for accepting the book in his series and Anthony Stavrianakis, who did a wonderful translation. Special thanks to Anne-Christine Taylor, Denis Vidal, Erik Bullot, Vinciane Despret, John Tresch, Chihiro Minato and Grégoire Mayor; to Graham Jones and Mikael Fischer for their encouragement and support and for convincing me that the book was worth translating; to Jacob Copeman, Willard McCarthy and Rahul Srivastava for their insights on the English version; and to all the people who gave me the opportunity to show and discuss this work on many occasions, organising workshops or screenings, enlightening me with their critical comments.

1

'Ganesh Yourself'

The Incarnation Game

The game we called *Ganesh Yourself* was born from an unusual wager. The idea was to design a machine that would allow anyone to put themselves in the place of the god Ganesh, giving this half-man, half-elephant deity a new way of existing, a new opportunity to fasten itself to this world, to grab on to something valued, and even to be glued to a device. Initially conceived as an experiment in the 'psychology of religion', the adventure quickly went beyond this, becoming a truly political experiment in the deconstruction not only of gods but also of machines (or at least of our faith in them), exploring unknown territories of the technological wild.[1] To what extent can an invisible entity enter a machine? Can a machine make a good metaphysical trap? Can you 'hack' a god, and what might be the virtues of such an exercise in piracy for those who engage in it? To incarnate Ganesh, anyone could apply, but whoever volunteered knew that they would be engaging in a risky exercise, as anyone could come to see them.

The system, tele-operated by a human, which we called an 'incarnant', and in which the only truly robotic component is, in fact, his trunk, animated by sensors, is at first glance quite close to Philip K. Dick's 'walkie-talkie', the idea of a portable god or system that allows a divine entity to teleport anywhere. Dick came up with this brilliant formula: 'God is the only theological entity that can be found everywhere, in the gutter and in the stars.' In *The Exegesis*, his several-thousand-page-long diary, Dick takes a malicious pleasure in speculating on its forms and imagines a world that is none other than our own, where 'every man would carry a bit of God inside him like a walkie-talkie (and much much more)'.[2] And he wonders: what if God were a kind of 'great system of artificial intelligence' that is profoundly active, that constantly goes off the rails but always lands on its feet? Or a 'form of plasma saturated with information, alive and intelligent', that would spend its time sending us encrypted signals via a whole load of information channels?

Determined not to leave it to theology to imagine transcendent entities, Dick engaged in one of the strangest undertakings of contemporary science fiction, looking everywhere for the activity of this ever more elusive entity, from the gutters to advertising slogans. Watching for the new faces that God could adopt, in a world increasingly overflowing with all kinds of technologies, he raises a problem of contemporary anthropological bearing: what monstrous forms could the absolute take on as our world continues to evolve? Can we imagine other ways of shaping it, and what other resources do we have to dialogue with the subterranean forces of the environment that surrounds us, its 'unknown bodies'?

It was in order to take up this delicate problem of civilization that we went to India and designed, with the complicity of the artist Zaven Paré, the robot named Bappa. Programming an autonomous simulacrum without an operator would have been time consuming and expensive, and certainly not as relevant as this interface, which is entirely dependent on a human incarnant in order to function, but which gives people the opportunity to put themselves in God's shoes and experience for themselves the difficulty of speaking as God. Anyone could apply and see their face and voice transmitted through a webcam and an internal overhead projector inside the robot. The robot looked like Ganesh, with its elephant trunk and ears. The operator could be positioned close by or at a distance, within the limits of the length of the cable that connected the headset to the central unit. Does a machine make a good *host entity*, a human a good mediator and a god a good *parasitic entity*? Is a fusion between and amongst them possible, a kind of endosymbiosis, ectosymbiosis or mesosymbiosis[3] as biologists say, unless, that is, we have to think differently about their relations? In spite of its simplicity, the device was sufficiently composite, blurred, fragile, chaotic even, for such questions to be brought to the fore. Everyone will find here, I hope, elements of more universal scope to rethink on other bases not only gods and machines, but the non-human in general. The experimental set-up had the advantage of confronting us with a very concrete dilemma: could a god that took the form of a machine be accepted, or else, just as soon as it is incarnated in this all-too-human form, will it be transferred elsewhere, returning to its spectral life in the depths of the invisible? This question concerns just as much anthropology and theology as robotics, and, more broadly, the ecology of non-humans.

The question of how to invent the right experimental apparatus to make a recalcitrant, unstable, elusive entity speak admits of very diverse answers depending on the context in which one is conducting the inquiry. Designing a 'god trap' in India has its own difficulties but also its advantages. In a polytheistic context, entities have the particularity of having a multitude of forms: flexible, sometimes fixed, sometimes mobile, they are willingly thought of as a *cornucopia of entities*, avatars etc., allowing for more abundant games of incarnation than monotheistic religions would allow.

It is not always easy to grasp where the threshold of the incarnatable lies in such a context, or the limit not to be crossed, beyond which a support is rejected as useless, bad or downright blasphemous. But it is precisely this limit that the experimental set-up of *Ganesh Yourself* made it possible to test, to provoke or to stimulate, like an energy point in acupuncture. If God emerges here only when certain conditions are met, there is no a priori reason why it should not be summoned into a machine, if the apparatus complies with the requirements of convocation (or invocation) that the entity needs in order to manifest itself. Many, among the incarnants, as well as their interlocutors, were sceptical from the start, but this did not prevent them from putting the device to the test, and to test their capacity to incarnate the voice of God, giving us the concrete opportunity to grasp, in practice, the conditions under which effects of divinity can occur.

It would be a mistake to see here only an experiment in implementing a god machine, an experiment whose aim would be to evaluate its conditions of viability, as if machines, ever more pervasive, were not already the object of a cult. Despite what the most reasonable among us think, the grafting of technology and religion has taken many and varied forms.[4] If Silicon Valley has today succumbed to compulsive forms of techno-millennialism and pours into the world its uninterrupted flow of transhumanist prophecies,[5] to do a genealogy of the 'divine sparks'[6] of machines would suppose going far back in time, and perhaps going to have a look elsewhere, as we propose to do here. Indeed, one impulse consists in absolutizing machines as a mode of redemption, but another, undoubtedly older and more widespread, has consisted in wanting to machine the absolute, either by conceiving, as anthropologists have already largely described, traps for *immaterial entities*[7] or even *occult machines*,[8] or else by imagining the cosmos in the form of a gigantic machine or by envisaging the gods themselves as *cosmic machines*.[9] Are there other possible ways, alternative relationships to the deification of machines that we have not yet even suspected? Can we imagine an experiment that helps us better define these unexplored possibilities and at the same time critique them?

An Exercise in Anthropology by Design

This exercise in experimental anthropology, or what we should rather call *anthropology by design*,[10] furthers the reflections started in a prior work, co-authored with Zaven Paré, on Japanese robotics.[11] It owes much to the collective work undertaken with Anne-Christine Taylor, Thierry Dufresne and Denis Vidal, which resulted in the exhibition *Persona*[12] concerning techniques of *entitification*, the multiple and various ways of personifying the components of the environment, of the milieu conceived of as living and acting, as well as the ways of knotting relations with non-humans, considered as persons. The work *Le Jour où les robots mangeront des pommes* (*The Day When Robots Will Eat Apples*) suggested in its own way a protocol that shifted away from experiments in ordinary robotics, an experiment that could be qualified as *empirical metaphysics* apropos of Hiroshi Ishiguro's Geminoid, a tele-operated humanoid robot that the Japanese roboticist conceived in his own image, designed to better understand the micro-movements of human interaction.

Our protocol consisted in playing to the full the machinic character of this game of incarnating a god, and exploring the range of possibilities in interaction, even the most improbable, experiences that would be propaedeutic to any radical critique of humanoid robotics. 'The time may come when for example a man may have to be restrained from attempting to rape a sewing machine,'[13] ironized Philip K. Dick, referencing Stanislaw Lem's infamous thought-experiment. Hence the importance of engaging in simulations, less in the mode of 'as if' or 'what would happen if?' than in the mode of '*My God, what if*...' With *Ganesh Yourself*, the logic of experimentation is pushed even further, especially since the experiment will not take place in the comfortable confines of a laboratory but in the tumultuous streets of Mumbai.

Anthropology is an experimental science unknown to itself. Hence it has never really formalized its modes of experimentation in the form of well-defined protocols. In experimental approaches to inquiry it is customary to progress step by step, to take things by their smallest constituents and then to increase in complexity. It is in order to find 'pure' conditions of experience that experimental psychology keeps its subjects in the laboratory and rarely ventures 'outside its walls', in situations where it cannot control all the parameters. Unfortunately, the human affairs that anthropologists deal with do not allow themselves to be manipulated with such ease, and are immediately grasped in all their exuberance. It is regrettable that when it claims to be 'experimental', anthropology often merely models itself on the protocols of neighbouring disciplines (experimental psychology and cognitive science in particular) in which it finds a reassuring rigour. *Ganesh Yourself* indicates another path to be taken, where the experiment is seen as an open game with vague hypotheses and invisible parameters, or what we could call *experimental systems with unknown variables*,[14] that we must then take up again through analysis. What is essential lies in the effects that this game provokes, the limits, the pitfalls and the otherwise unthinkable possibilities that it makes possible to test, as well as the preconceived ideas that it manages to displace.

In the experiment that follows, which has faint echoes of a Turing test, what is at stake for the gods is to learn to be machines, but perhaps also to be more in tune with human affairs. For the humans participating in the experiment, the crux is to attain the perfection of the gods, and for the machines, unaware of their ignorance, they will have to learn many things from humans and gods, if they would like us to worship them. *Ganesh Yourself* prefigures what could be a kind of 'deep mystic learning',[15] if such an appropriation might be permitted, which aims to determine how machines could learn what gods have long known, with the necessary complicity of humans.

The roboticist Masahiro Mori, author of *The Buddha in the Robot*,[16] considered the ways in which machines could go beyond the strict threshold of utility, to tie themselves to humans in relations full of spirituality. Machines devoid of consciousness incarnated, for Mori, the state of absolute serenity aimed at in the meditative practice of *zazen*. We can of course ask ourselves how such speculations were possible,[17] as well as why, at this conjuncture, rampant Silicon Valley engineers are thinking of programming a god endowed with artificial intelligence? Behind these occasionally provocative, but only seemingly original, ideas we must ask ourselves what pact we wish to make with machines and whether humbler experiments could not be conceived that would bring the fusion between the spiritual and the material within our reach, including the self-analysis of our deepest fantasies about it.

Prototypical robotics experiments usually take place within the confines of a laboratory, using questionnaires, just like standard experiments in the psychology of religion. In these experiments, the experimenter always takes a bird's-eye view, with questions defined in advance of any actual inquiry. Our intervention was to set forth an incarnation technology, which people could make use of as they wished, engaging in their own role-playing, inventing their own question-and-answer rituals and deciding for themselves which topics to address or which problems were worthy of being submitted to a god. And if we did not know at all how this new medium of divinity would be perceived in advance, one did not have to be a fortune teller to know that any pretence to being God was doomed to failure. We can thus, *alas*, cut the suspense.

Bappa was never really perceived as a plausible form of divinity without long debates, and was used instead as an interface to rethink divine affairs, by people determined to put themselves for a while in God's place and to switch roles (questioner/questioned), not letting themselves be dictated to by any imposed psychology of religion. We had prepared some phrases to give to the incarnant in case he or she had trouble playing the part, but we never used them. And if, to our great surprise, the device was quickly taken up as a plausible interlocutory interface, the ways in which it was appropriated were completely unexpected, diverting its uses, tearing up any preconceived objectives, according to subversive logics that will be examined in detail.

Beyond the Turing Test and the Ideology of Machines as 'Traps'

In a well-known article from 1950,[18] Alan Turing presented a fundamental transformation in how to conceptualize the design of 'intelligent' machines. Instead of asking the question 'Do machines think?', a question he considered absurd, Turing posed a more nuanced problem: under what conditions could a machine deceive us into thinking that it thinks? Turing suggests the following game inspired by the well-known 'imitation game' played at fairgrounds: an individual, player C, is in a room and exchanges only written notes with two other players, who cannot be seen: A, who is a man, and B, who is woman. Player A tries to trick C and player B answers questions only truthfully, trying to help C make the right decision about which of them is the man and the woman. Turing then asked, 'what will happen when a machine takes the part of A in this game?' Variations on the game include having both A and B try to confuse and trick the inquirer.

Turing's test was designed to generate as much ambiguity as possible in order to trick the questioner. It aimed to test the capacities of a machine to confuse us not only as to whether or not it is a machine or a human, but also as to whether it is a man or a woman. In other words, machines appear endowed with intentionality for Turing only if they deceive us about their true nature as machines and if they are able to get us to impute attributes to them which they do not in fact possess.

Tests similar to the one designed by Turing are set up spontaneously each time we have to come to a judgement about who or what we are dealing with, an answer to the question 'who is there?' and, if there is something there, an entity, an agent, then what kind of thing it is. We experience this every day: as machines become increasingly networked, the question of 'is there someone behind it' and, if so, 'who is there', increasingly becomes a problem. Behind one's computer screen, one may suspect (and one may be right), that there is a network, perhaps other entities, other machines, even other people. In the same way, in *Ganesh Yourself* we can consider that the problem is to know under which conditions someone can deceive us about the fact that he is a god, but in reality, the question is first to know 'who is there' or 'who we are dealing with', beyond the initial resemblance of a god that seems familiar to us.

Each time an interlocutor sits down with Bappa to talk, a kind of Turing test is set up. And it is then by going through an ontological test of classification ('Are you man, machine or God?') that *Ganesh Yourself* makes sense. Indeed, the dynamic of any interaction with Bappa is the will to test its possibilities, its potential to channel the deity. We will see how the participants in the experiment may even have unknowingly rendered Turing totally obsolete, or else, inadvertently proposed another, much less perverse test, a counter-test based not on the capacity of a machine to deceive its interlocutor but on its potential to house entities, as elusive and indefinable as they may be, to make present something other than itself, from the most visible to the most invisible: a kind of test to evaluate the spectral possibilities of things, of the capacity of machines to make good mediums, even for an interface as rudimentary as ours.

To what extent can an invisible entity possess a body? To what extent can matter crystallize forces that overwhelm it? These are problems that admit of very diverse answers depending on the context in which anthropologists choose to situate themselves. Being in India will help because of the fecundity of the Hindu soil with respect to these questions. The reader will not be surprised that Gilles Deleuze, expressing his curiosity for these other forces that could be connected to silicon (synonymous as it is with the taxonomic kingdom of machines), invites us to look at this corner of the world:

> I believe that there will be, that one could find, forms which in effect will be no longer God, nor human, nor superhuman, but which will be extraordinary vegetal or animal forms ... It is India that interests me above all in this regard.[19]

The prolific pantheon of anthropomorphic gods in modern Hinduism has little to do with the aniconic religion of Vedic India.[20] The personification of forces has thus not always taken the form of an anthropomorphic bestiary and ritual alchemy, that concrete science of composites, so as to determine which powers, which entities to cultivate, in what way and for what effects. And in spite of the appearance of stability, it has also evolved a lot. Despite those who, like Hegel, have rejected fantastic art as a way to access the absolute,[21] the Hindu pantheon as anthropologists observe it today is more anthropomorphic than ever, and highly elastic: the gods spend their time incarnating themselves, metamorphosing, devoting themselves to feeling and perceiving in multiple ways, and have to answer when they are invoked, at the same time as humans have fun cobbling together new supports, shaking up the limits of what is possible to experience in terms of the effects of the presence of the divine in everyday life.

Bappa will allow us to experiment very concretely with these effects of presence. On arriving in Mumbai we had to request the services of an officiant to bless the robot. Although the *paṇḍit* had no objection to subjecting Bappa to a ritual to 'open the eye' like any other idol, the status of this tele-presence interface was not settled. A casting session was then organized to recruit incarnants, and after a few weeks, despite (or perhaps because of) its ambiguous status, the machine was inundated with applicants, overwhelmed by its success.

In a country where speculating about the nature of God is a widespread pastime, people were eager to play the game, debating for hours on end, via the robot, unresolved theological questions or problems that concerned them in their daily lives (pollution, corruption and so on). Priests had no problem integrating the machine into their own set of props for the transmission of their mantras. Political activists, including environmentalists, gays and feminists, made use of the interface to convey messages of social reform, realizing that technology with a divine face was a good way to make themselves heard. And so the idea was born to open a consulting room in the city called 'Ganesh Yourself', a mobile office that we could take anywhere. Astrologers took turns giving consultations, realizing that using a tele-presence interface allowed for a different type of dialogue with their clients. And while not everyone made for a satisfactory god, Bappa offered a rare opportunity to have, at best, a dialogue with God in the mode of an ordinary conversation, or else, at worst, to expose the imposter who tried to play him in such a crude way.

It is likely that the experiment would have taken a different turn elsewhere, but one should not imagine that in India the game seemed easy. It is true that in a polytheistic context there is no reason to refuse to summon a deity into a new medium, if the latter complies with the requirements of convocation (or invocation) that the entity needs in order to manifest itself.[22] Just as two voices combine in polyphony to produce a third, the phenomena of emergence respond to a simple formula: 1 + 1 = 3. If we follow this formula, is a human plus a machine enough to make a god emerge? Nothing prevents us from imagining it theoretically, but in practice, it is not certain that the divine is better obtained by addition than by subtraction.

Machining the Invisible: A Concrete Way for Metaphysical Inquiry

It wouldn't be too much of an exaggeration to say that among anthropological theories of religion, the most common envisage gods as psychic creations, 'inventions', 'optical illusions', 'mind traps'[23] or 'lures',[24] translating the remarkable imagination by which humans have shown themselves to be able to orient themselves and to draw on such beneficial powers in an uncontrollable environment populated with hostile forces. While the trap model is useful for thinking about a host of things, presupposing the trifold division between a lure, a lured and someone or something that lures, it cannot be applied to gods entirely satisfactorily, for we would have to determine how traps are different from works of art, which, when they intrigue us, can become real metaphysical traps,[25] or even how they are different from algorithms in the digital age, which have the task of recording all traces of our actions so as to return to us in the form of advertising recommendations.[26] *Ganesh Yourself* will perhaps make it possible to move forward in this field of inquiry by asking to what extent the invisible is (or is not) machinable, or under what conditions something of the absolute[27] can be taken up within a machine and through what mechanisms a divine feeling can emerge.

Unlike the Buddha, who invites serenity and meditation at once, and in whom Mori sees a model for the spiritual machines of the future, the god Ganesh is zoomorphic and is striking in his constitutive anomaly. It is with the utmost seriousness that the Indian Prime Minister, Narendra Modi, declared before a gathering of doctors and health professionals: 'We can be proud of what our country has achieved in medical science in ancient times ... We worship the god Ganesha. A surgeon of that time had thus grafted the head of an elephant onto the body of a human being and invented the practice of plastic surgery' (25 October 2014).

Why in Hindu polytheism have such entities taken such a form, embracing such a field of action? Some people are still trying to solve the enigma of Ganesh's original deformity, mobilizing very curious theories that range from purely symbolic operations through to grafts, or even a deified genetic anomaly. *Paṇḍit* Lachmidhar Shastri once suggested that Ganesh was the result of a *suggestion* from the environment rather than a conscious effort on the part of an artist.[28] He does not explain, however, how the phenomenon of induction actually occurred: an elephant talking into a human's ear? An individual imitating an elephant? A man born with a trunk? A creature that appeared in the dream of a possessed person? An indigenous Amazonian would undoubtedly have no difficulty in accepting the thesis of appearance by 'suggestion', where dream entities have their place in the ecology of beings. But in India, other hypotheses have been formulated, notably the idea that the elephant and the rat could have been totemic animals, emblematic of two dynasties that clashed at the time of Alexander, the victor adopting the elephant rather than the rat as an emblematic entity. Is the hypothesis of a chimerical composition, a deliberate act of political invention, more plausible than a human with the head of an elephant, walking in the wild or else the first trans-species cosmetic surgery?[29]

In *The Elephant Man and Other Reminiscences* (1923), the English surgeon Frederick Treves, who dealt with the famous case of John Merrick, who became a fairground attraction, compares his patient to the well-known Hindu deity. This line of thought was then widely taken up by novelists[30] who had fun exploring the hypothesis that John Merrick could have been an incarnation of this unjustly forgotten god.

The thesis of a monstrosity made god goes against the rejection of which Merrick was victim in a Victorian England otherwise little inclined to accept deformity. There are many cases in present-day India that demonstrate that a resemblance to divinity at birth has a good chance of being interpreted as a divine manifestation and of providing a destiny for the bearer of this physical characteristic. The controversies of the historians, and the exuberance of science fiction novelists who have taken up the theme, make this never sufficiently lively entity live in its own way. Considering this lack as a driving force and Ganesh as an entity never truly finished, an unfinished experiment without a true author, is a first step towards the recognition of its fundamental plasticity.

In a universe that Whitehead liked to imagine as 'a dim background shot across by isolated vivid effects charged with emotional excitements',[31] Ganesh seems to have more or less proven himself as a *power* to protect against calamity and attract prosperity (he is nicknamed the destroyer of obstacles, *vighnahartā*),[32] and many people place an idol of the elephant god in one place or another in the home to prevent negative waves from entering. Ganesh is not, however, a simple entity. He conceals a *myriad of forms*, a whole matrix, composed of solid, animated forms, some forged by the hand of man, others temporary incarnations according to the circumstances, and others so-called 'self-generated' forms that have no creator (*svayambhū*).[33]

Anthropologists endeavour to work out the regime of life proper to such entities by examining first of all the mythologies and rituals that are attached to them, yet they rarely dare to put into circulation a new form that breaks with the usual routines. If *Ganesh Yourself* is an offbeat experiment it is because it was not at all certain, when we conceived it, that Ganesh would take to it. In the same way that the conception of a humanoid endowed with the faculty of vision, or the capacity to walk, obliges the roboticist to start from scratch all the physiology of the eye or of bipedalism, to make a god also produces an effect of reinitialization, which it will be necessary to try to better qualify. Can we restart the field of invisible forces, the mechanisms of belief, or what James called *the will to believe*? Why would a machine tele-operated by a human be a good way to operate a radical critique of what is potentially cultic in our relationship to machines? All this may seem quite abstract at the moment, but it will become very concrete as we give more detailed accounts of the experiment.

Conceiving a god that speaks or answers, produces, amongst other things, a break with the more or less implicit contract that characterizes the ordinary devotional relationship. The intimate dialogue with an inner voice is externalized in the form of a dialogue spoken aloud. And insofar as we have multiplied the incarnations, thanks to the capacity for speech that is here freed up, we will see loom up a crowd of contenders to divinity, but also a multitude of unquestioned elements, notions, collectives and buried desires that rarely find expression. Such an experiment is not without the inevitable triggering of a cascade of questions, both ethical and political. The Japanese roboticist Hiroshi Ishiguro faced a similar problem when he designed his Geminoid, a humanoid robot made in his image.[34]

The Geminoid, which was supposed to allow roboticists to study micromovements in face-to-face human interaction, was also a new creature, a new avatar, intended to be used far beyond the confines of his laboratory, fanning fears and fantasies. Ishiguro had to accept the two-headed aspect of his creature, both a tool of knowledge and a provocation destined to precipitate phenomena that he could not foresee. However, it is all its accumulated uses (scientific, artistic, utopian) that make the Geminoid more than a simple robotic experiment, a *limit-experience* and even an experiment in civilization, which is as much a matter of science and of art as of politics. With much smaller means, *Ganesh Yourself* raises similar questions but also problems of its own which will become clearer throughout the book.

Ganeshology: Why a Hybrid Makes a Good Speculative Device

It is regrettable that invisible entities, and gods in particular, have not been the subject of as elaborate a natural history as machines. Jacques Lafitte, for example, some time ago, bade a look through the history of machines, like a zoologist observing the emergence of a species in the process of development, giving an account of the formation of lineages, moments of mutation, mechanisms of speciation and hybridization.[35] Simondon did the same, resituating the emergence of machines in the history of the living, including amoebas just as much as computers, tracing the exceptional heterogeneity of their modes of individuation.[36]

From machine tools to cybernetic machines that increasingly hide their mechanisms, endowing themselves with programmed interiorities, all the way to the network-entities of the computer age that have little to do with their ancestors, the concept of a machine has not stopped changing, adopting evermore complex forms, shaking up the relation between inside and outside. By dint of having considered the gods as immutable powers, insensitive to their slow regime of fluctuation, we forgot to grant them the right to evolve, and thus to include them in natural history. Deceived by the anthropomorphic aspect they could take on in ancient paganism or in late Hinduism, we have largely underestimated the fact that these energetic bodies are perhaps among the first circuits ever imagined, that they are like real, energy-consuming machines (especially when they have to be fed with offerings as in India) and that they have never stopped mutating, acquiring new properties, being abandoned in favour of others better adapted to their environment, having a better chance of persisting.

Bappa could be seen as a rope 'fastened between the animal and the superhuman', as Nietzsche said, over an abyss that is perhaps nothing more than our present and its latent possibilities. This seemingly divine gadget or simulacrum of the god Ganesh is in fact a machine of divine deconstruction and might prefigure other forms, another species, another pact with animals and machines. Bappa extends the history of animals, and gods as much as that of machines, in a kind of spiritual hybrid, but it aims above all to question this history and the very instability of the possible forms of deification, operating a kind of ontological rethinking as to the choice of the most suitable form. What is worthy of worship in a given collective and at a given time does not always remain unthought, and can be the object of intense speculation for those concerned. The compromises adopted, more or less durable in the form of this or that cult, can suddenly crack under the chaotic and subterranean pressures of recasting and overhaul, which lead to questioning the current forms of deification in order to imagine other forms that are more adapted or are simply more *alive*.[37]

In a world that does not stop evolving and disconcerting us, what can a god do, what can machines do and what can a new interface for deification provide? Everyone in India (or almost everyone) has an opinion about gods, their scope of action and their sphere of responsibility. We will see people engaging in long debates through and by way of the machine, trying to sort out 'live hypotheses' from 'dead hypotheses' in William James's sense.[38] People are constantly doing their own psychology of religion and their own metaphysics. *Ganesh Yourself* echoes this. The interaction with Bappa is first and foremost an 'analytical' moment, where critical thinking, doubt and uncertainty are not stifled but, on the contrary, constitute the very fuel of the interaction, as does the ability of beings to rethink the situations in which they are involved.

This speculative activity is all the richer in a context where the gods do not really have a stable body, where they can be embodied in idols but also abstracted from them and disappear into the depths of the cosmos. Speculating on the nature of the divine becomes, in this context, a pervasive pastime. It is not uncommon in India to be told that a god can be there at any time, incarnate itself in an idol, in a stone, and even take human form, but also that God is a unique principle, a pure abstraction, unrepresentable, that he will always surpass any attempt at materialization, or that he is the embodiment of an impossible point of view.

To have even the slightest chance of understanding what is going on in this more or less thick layer of reality, at the same time extremely populated and paradoxical, it would be necessary to hold off using the very notion of 'god', to suspend it as long as possible. And this is perhaps one of the advantages of our experiential and experimental set-up: past the vague impression of resemblance with Ganesh, which does not fool anybody, something of the order of the absolute allows itself to be felt in this boundary-presence, but in a deeply whimsical mood. It is these apparently uncontrollable workings, which, as we shall see, are largely interactive, that we hope to better grasp.

Anthropologists of India will have no trouble imagining *The Incarnation Game* taking place in the streets of Mumbai. They have spent their time describing highly complex processes of manifestation with gods who are often unpredictable, endowed with capacities of movement that surpass human comprehension;[39] gods who must be summoned by games of questions and answers,[40] in the course of ritualized interactions in which they are forced to express themselves by signals of presence that must then be interpreted. It is not surprising that, within Hindu contexts, anthropologists have so heavily debated the right way to 'classify the gods'[41] amidst their profuse manifestations, but also how to make them speak, and even how to insult them, on occasion, when they prove ineffective or counterproductive.[42]

Polytheisms, from here and elsewhere, have not only abounded in various supports whose divinity-effects[43] must be evaluated, but also in trials and tests of all kinds[44] that aim to put to the test, from the point of view of the devotee himself, the pact he maintains with the gods. Under what conditions can a new composite produce a 'power' that is acceptable? This problem will take on a fundamentally novel form with the robot Ganesh, considering that it is a machine and therefore a new possibility for incarnation.

Any so-called 'intelligent' machine or tele-operated system raises implicit theological questions, as Norbert Wiener, the founder of cybernetics, already remarked in his book *God & Golem*.[45] Unfortunately, it is almost always from our monotheistic traditions and from a Judeo-Christian paradigm that the convergences between religion and technology are approached.[46] Shifting to India has the virtue of aiding us in considering the problem in a different way. What does it take for us to have the impression, during an encounter with a machine, that an entity much larger and more mysterious than its medium is indeed present, alive? Is it at the moment when 'a rightness is attained or missed, with more or less completeness of attainment or omission',[47] as Whitehead invites us to think? Whitehead's God is an 'actual entity', always endowed with two sides, one turned towards eternity and the other towards the contingency of the present; he can give himself to being apprehended in an occasional experience, in the form of a sensation or an effect. At the same time actual and virtual, it is fundamentally bipolar.[48]

In a conversation with Ganesh, not only will there be an oscillation between two poles, between the abstract and the concrete, the immortal and the actual, or between the actual and the virtual, but in an even more vertiginous way, amongst and between the human, the machine, a multitude of potentially singular gods and all the varieties of their relations. We can only grasp what is at stake in this very plastic zone of incarnation by letting the experiment impose its effectuations on us. The gamble of *Ganesh Yourself* is that the most important thing is to play up, to the greatest extent, the disturbance that the game of incarnation provokes. It is eminently subjective and palpable, for those who thought they could play along 'as if', or who thought that 'it's all just a game', brimming with the bugs of an imperfect machine, or else when the incarnating person realizes that his or her voice may only partially be his or her own, and that it comes from elsewhere. It is there, in the hollows of a process of incarnation that we thought illusory or doomed to failure, that the multitude of possible gods will be best felt, including those for which we have not opted, those we could have adopted or those we could make our own.

Notes

1. On non-linear ways to look at the history of technology and wild tech as a driving force escaping the simplistic alternative between the high and the low, see Emmanuel Grimaud, Yann-Philippe Tastevin and Denis Vidal (eds), *Low Tech? Wild Tech!* Techniques et Culture 67, 2017.
2. Philip K. Dick, *The Exegesis of Philip K. Dick*, ed. Pamela Jackson and Jonathan Lethem (New York: Houghton Mifflin Harcourt, 2011), 167.

3. On the panoply of possibilities for symbiosis at the scale of life, see Lynn Margulis and Dorion Sagan, *Microcosmos: Four Billion Years of Evolution from Our Microbial Ancestors* (Berkeley: University of California Press, 1986).
4. For examples from ancient times, see Lucien de Samosate, *Alexandre ou le faux prophète*, vol. I (Paris: Hachette, 2002), which gives a good overview of the profusion of mechanisms in ancient paganism. Some oracular devices dissociate the object from the voice of the operator. This is the case, for example, of the Glycon, a serpent deity ventriloquized by the prophet Alexander of Abonuteichos, who flooded Rome with his cryptic messages and predictions in the second century.
5. The Technological Singularity is one such prophecy. Vernor Vinge, 'The coming Technological Singularity: How to survive in the post-human era', *Whole Earth Review*, 1993. And on its later transhumanist versions, see Eliezer S. Yudkowsky, *The Singularitarian Principles*, Version 1.0.2, Compact edition, 2000; Max More, 'Principles of Extropy (Version 3.11): An evolving framework of values and standards for continuously improving the human condition', *Extropy Institute*, 2003; Ray Kurzweil, *The Singularity is Near. When Humans Transcend Biology* (New York: Viking Books), 2005. For a fun inquiry amongst transhumanists, see Mark O'Connell, *To Be A Machine: Adventures Among Cyborgs, Utopians, Hackers, and the Futurists Solving the Modest Problem of Death* (New York: Doubleday, 2017); Abou Farman's excellent analysis 'Re-enchantment cosmologies: mastery and obsolescence in an intelligent universe', *Anthropological Quarterly*, no. 4 (2012): 1069–88; as well as Gabriel Dorthe's terrific article, 'L'Immortalité et ses impatients', *Socio-anthropologie*, no. 31 (2015): 127–38.
6. The expression is taken from Erik Davis, *TechGnosis. Myth, Magic and Mysticism in the Age of Information* (London: Serpent's Tail, 1999).
7. The genealogy of traps to capture invisible powers is, in large part, yet to be done. Dogon masks could have been envisaged by Marcel Griaule as techniques of trapping immaterial principles. Marcel Griaule, 'Cours sur les techniques' (11 March 1953), 2–3, in *Méthode de l'ethnographie* (Paris: Presses universitaires de France, 1957).
8. Spiritism abounds in immaterial traps, see Philippe Baudouin, 'Archéologie des machines occultes', *Terrain*, no. 69 (2018): 96–113.
9. On the Hindu gods as cosmic machines and Ganesha in particular as a factory for treating misfortunes, see Grimaud, *Dieux et robots*.
10. Over the last decade the term 'experimental' has been increasingly used to characterize work in anthropology. Cf. Paul Rabinow and Gaymon Bennett, *Designing Human Practices: An Experiment with Synthetic Biology* (Chicago: University of Chicago Press, 2012); Michael Fischer, *Anthropology in the Meantime: Experimental Ethnography, Theory, and Method for the Twenty-First Century* (Durham: Duke University Press, 2018); Adolfo Estalella and Tomás Sánchez Criado (eds), *Experimental Collaborations. Ethnography Through Fieldwork Devices* (New York, Oxford: Berghahn, 2018). For a discussion of this polysemic word and the various aspects of anthropological experimentation, see Emmanuel Grimaud and Anthony Stavrianakis, 'We have never been not experimental/how much more experimental can anthropology be?' *Annual Reviews of Anthropology*, forthcoming. On the relation between anthropology and design, see Keith Murphy, *Designs and Anthropologies: Frictions and Affinities* (Santa Fe: School for

Advanced Research Press, 2021). On ethnographic invention in general and the notion of 'the device' in particular, see Tomás Criado and Adolfo Estalella (eds), *An Ethnographic Inventory: Field Devices for Anthropological Inquiry* (London: Routledge, 2023).

11. Emmanuel Grimaud and Zaven Paré, *Le Jour où les robots mangeront des pommes. Conversations avec un Geminoid* (Paris: Petra, 2012).
12. Thierry Dufrene, Emmanuel Grimaud, Denis Vidal and Anne-Christine Taylor (eds), *Persona, étrangement humain* (Paris: Musée du quai Branly/ Actes Sud, 2016).
13. Philip K. Dick, 'The android and the human' (1972) in Robert Latham (ed.), *Science Fiction Criticism: An Anthology of Essential Writings* (London: Bloomsbury, 2017), 296.
14. 'Experimental systems are vehicles for materialising questions', Hans-Jorg Rheinberger, *Toward a History of Epistemic Things: Synthesizing Proteins in the Test Tube* (Stanford, CA: Stanford University Press, 1997), 28.
15. Deep learning is a form of programming now familiar to engineers, which aims to train machines to learn by themselves.
16. Masahiro Mori, *The Buddha in the Robot: A Robot Engineer's Thoughts on Science and Religion* (Tokyo: Rissho Kosei-kai, 1981).
17. Japanese roboticists are not the only ones to have speculated on the subject, combining Buddhism and robotics. On the Indian use of religious automata and Hindu conceptions of cosmic machines, see Grimaud, *Dieux et robots.*
18. Alan Mathison Turing, 'Computing machinery and intelligence', *Mind*, no. 59 (1950): 433–60.
19. Gilles Deleuze, 'Foucault', Seminar at Paris 8 University, 8 April 1986, https://deleuze.cla.purdue.edu/seminars/foucault/lecture-18, translation modified.
20. Charles Malamoud has stressed this point, notably in *Cooking the World: Ritual and Thought in Ancient India*, trans., David White (Oxford: Oxford University Press, 1996 [1989]) and in *La Danse des pierres. Études sur la scène sacrificielle dans l'Inde ancienne* (Paris: Seuil, 2005).
21. If there is one author who seems to have violently rejected the idea that one can reach the absolute in this way, it is Hegel, who sees in Indian art the expression of an excessive taste for distortion. In Indian art, 'the imagination thus goes from one extreme to another, rising very high only to fall even lower, wandering without support, without guide and without goal, in a world of representations at once grandiose, bizarre and grotesque'. Charles Bénard, *Sur l'esthétique de Hegel* (Paris: Ladrange, 1852), 270. And in Hegel's own words about the Indian imagination: 'the fantasticalness of a restless inquisitive imagination and an unintelligent talent for configuration'. G. W. F. Hegel, trans. T. M. Knox, *Hegel's Aesthetics: Lectures on Fine Art* (Oxford: Clarendon Press, 1975), 340. For Hegel, God could only be elsewhere, before us, at the end of history, at the culminating point of the development of the Spirit. Cf. Mark Alizart, *Informatique celeste* (Paris: Presses universitaires de France, 2018).
22. Tracing the emergence of monotheism as the religion of a jealous god, Jan Assman points out that in ancient polytheisms (from Mesopotamia to Greece, passing via Egypt), when one encountered gods, myths and rituals that differed from one's own, the reflex was to infer the existence of a common divine reality that manifested itself in a plurality of different appearances rather than to condemn these manifestations as false. See J. Assman, *Moses*

the Egyptian: The Memory of Egypt in Western Monotheism (Boston: Harvard University Press, 1998) and *The Price of Monotheism*, trans. Robert Savage (Stanford: Stanford University Press, 2010). Maurizio Bettini sees in this principle of tolerance a good reason to praise polytheism, *Éloge du polythéisme* (Paris: Les Belles Lettres, 2016).

23. On the notion of mind traps (pièges à penser), see Pascal Boyer, *Religion Explained: The Human Instincts that Fashion Gods, Spirits and Ancestors* (New York: Basic Books, 2001).
24. André Leroi-Gourhan defines the lure as an 'organ of seduction always supplemented by a murderous organ', in *Milieu et Techniques* (Paris: Albin Michel, 1973), 76.
25. Alfred Gell, 'Vogel's net: Traps as artworks and artworks as traps', *Journal of Material Culture*, no. 1 (1996): 15–38.
26. Nick Seaver provides a fine analysis of the anthropological literature on traps before turning his attention to captology in the digital age, 'Captivating algorithms: Recommender systems as traps', *Journal of Material Culture*, no. 24(4) (2019): 421-36.
27. On the experiential modes developed by Hindu mysticism to access the absolute (*brāhmaṇ*), understood as cosmic consciousness irreducible to any form, see Michel Hulin, *Comment la philosophie indienne s'est développée: la querelle brahmanes-bouddhistes* (Paris: Éditions du Panama, 2008), and, above all, *La Mystique sauvage, aux antipodes de l'esprit* (Paris: Presses universitaires de France, 1993).
28. In *Proceedings and Transactions of the Ninth All-India Oriental Conference*, Trivandrum, 1937, cited in S. M. Michael, 'The origin of the Ganapati Cult', *Asian Folklore Studies*, no. 42(1) (1983): 91–116.
29. Ganesh is undoubtedly one of the entities of the Hindu pantheon to have lent itself to every metamorphosis. It was not uncommon, for example, to find portraits of Gandhi with an elephant trunk during the struggle for Independence. See, on this point, Raminder Kaur, *Performative Politics and the Cultures of Hinduism: Public Uses of Religion in Western India* (London: Anthem South Asian Studies, 2003).
30. See, for example, the science-fiction novel by Xavier Mauméjean, *Ganesha, Mémoires de l'homme-éléphant* (Paris: Mnemos Editions, 2007).
31. Alfred North Whitehead, *Religion in the Making: Lowell Lectures* (Cambridge: Cambridge University Press, 1926), 14.
32. S. M. Michael suggests that there was a time where Ganesha was rather likened to a demon, creator of catastrophes. Michael, 'The origin of the Ganapati Cult', Asian Folklore Studies, no. 42(1) (1983): 91–16.
33. On the multitude of modelling experiments to which it gives rise among artisans during the *Gaṇeśa Chaturthī* celebrations, see Grimaud, *Dieux et robots*.
34. Grimaud and Paré, *Le Jour où les robots mangeront des pommes*.
35. Jacques Lafitte, *Réflexions sur la science des machines* (Paris: Vrin, 1972 [1932]).
36. Georges Simondon, *Du mode d'existence des objets techniques* (Paris: Aubier, 1958).
37. It is enough to see how, in our own worlds, Gaia as a cosmic organism is currently the object of very diverse attempts at deification or how new age cults are reviving various forms of polytheism. Environmental theology has been

asking since at least the 1960s which religious system is best suited to the ecological crisis. See Lynn Townsend White Jr, 'The historical roots of our ecological crisis', *Science*, no. 155 (1967): 1203–7; or M. E. Tucker and J. A. Grim (eds), *Worldviews and Ecology* (New York: Orbis Books, 1994).

38. A live hypothesis 'is one which appeals as a real possibility to him to whom it is proposed ... The maximum of liveness in an hypothesis means willingness to act irrevocably.' A dead hypothesis, by contrast, 'refuses to scintillate with any credibility at all'. William James, *The Will to Believe and Other Essays in Popular Philosophy* (New York: Longmans, Green and Co., 1897), 2–3.
39. On this point see Denis Vidal *Le Culte des divinités locales dans une région d'Himachal Pradesh* (Paris: ORSTOM, 1988).
40. A very nice example of this type of the questioning to which divinities are sometimes subjected is found in Daniela Berti, *La Parole des Dieux* (Paris: CNRS, 2001).
41. Véronique Bouillier and Gérard Toffin (eds), *Classer les dieux?* Collection Purushartha, vol. 15 (Paris: EHESS, 1993).
42. A nice overview of why goddesses are insulted can be found in Olivier Herrenschmidt, 'Choisir ses dieux', in *Persona, étrangement humain* (Paris: Actes Sud, 2016), 214–15, which describes how goddesses, accused of bringing calamities to the villages of Andhra Pradhesh, are insulted.
43. On the notion of simulacra in ancient Greece and what Christianity owes to the Epicurean conception of the divinized man, see Renée Koch, *Comment peut-on être dieu, la secte d'Épicure* (Paris: Belin, 2005).
44. On the specificity of interactional games with non-humans in general, robots and gods in particular, see Denis Vidal's *Frontières de l'humain. Dieux, figures de cire, robots, et autres artefacts* (Paris: Alma, 2016).
45. 'There are at least three points in cybernetics which appear to me to be relevant to religious issues. One of these concerns machines which learn; one concerns machines which reproduce themselves; and one the coordination of machine and man.' Norbert Wiener, *God & Golem, Inc. A Comment on Certain Points Where Cybernetics Impinges on Religion* (Cambridge, MA: MIT Press, 1964), 11.
46. On the collision between computer science and theology, see Marc Alizart, *Informatique céleste*. On the use of the Apocalypse in the techno-prophecies of contemporary transhumanists, see E. Grimaud, 'Silicon prophecy. The Technological Singularity according to Vernor Vinge', *Terrain*, no. 71 (2019).
47. Whitehead, *Religion in the Making*, 50.
48. Alfred North Whitehead, *Process and Reality; An Essay in Cosmology. Gifford Lectures 1927–1928*, corrected edition, ed. D. R. Griffin and D. W. Sherburne (New York: The Free Press, [1978] 1985).

2

Divine (Im)Posture

Everyone has their own idea of the voice and mission of God. As the first incarnation, Karan's strategy was to take on a divine posture from the get-go, rather than to patiently work towards it. We had recruited him through a classified ad. He was twenty-six years old, from Gwalior, and had a degree in commerce. Rather shy, with a soft voice, he had, at first sight, no particularly distinguishing features, but when he closed his eyes and went into a state of meditation, there was no doubt in his mind that he was Ganesha, and that is the reason why he was chosen. 'Ganesha is in me,' he said. Whenever he was about to embody Ganesha, he would do some breathing exercises to ready himself. He would repeat the syllable *Oṃ*, or else the phrase *Oṃ namaḥ śivāya*, over and over. His guttural sounds and the breath his respiration produced drew people toward the robot, intrigued, as they lingered in front of it for longer or shorter periods.

Karan had a plan, but we didn't know it. He had decided that he would not end up immersed in the sea, a fate reserved for all idols of Ganesha once the ritual festivities were over. During his first interaction with children, this was the first thing he said to them, as if trying to convince himself: 'Unlike all the idols, I will not be immersed. Why not? Because I am an electronic Ganesh.' No one had whispered such words to him. Karan had identified with the machine; he had incorporated it into his being to such an extent that he had decided that their destiny would not be sealed by the words of the *paṇḍit*. 'I have come to Earth with a mission. My purpose concerns you all. Devotion to the god Ganesh must grow. I will travel the entire city for this purpose alone. I am Ganesha, the son of Gauri. And you may ask me questions.'

To the question 'Where are you from?' Karan had an answer ready. 'I am the son of our great god Shankar. My father has many names. I came down from Mount Kailash to be with you. My parameters have been set through a rite to open the eyes [*prāṅa pratiṣṭha*]. What is the total number of gods and goddesses? Thirty-three thousand, and I am one of them!' And when a child asked him the question, 'Where has your mount gone? What have you done with your rat?' his improvised answer was curious: 'As you know, I lost my mount. I am trying to find it. I practise meditation every day and then I talk to people. I am looking for the rat that I use as a mount during my meditation. He is around here somewhere. Do you know what meditation consists of?' As the answer was negative, Karan felt compelled to explain, 'Meditation is connecting to your inner self and we see another world. Everyone can practise meditation. Because you become very strong like this, you should all practise meditation. Repeat after me: *Oṃ namaḥ śivāya* ...'

We were in a fishing village by the Bay. The village had the appearance of a slum. A few metres away, people were using the beach to relieve themselves. Meditation was a pastime of the wealthy classes, not so much a concern of the fishermen. But Karan was there to convert them, to educate them to his own religion, that of the great Brahmanic gods, a form of religiosity based today on two pillars: ritual and meditation.

Thanks to Karan, we knew that *Ganesh Yourself* would not be just a game. The danger was to succumb to a robotically assisted evangelism, or that all interaction would get bogged down in a mythological quiz where the god is the teacher of good behaviour, or even to stir up class warfare through the gods. Karan was not afraid of any of these pitfalls. He plunged cheerfully into all the traps that were set for him. And we let him say and do. To seek confrontation was risky. Either he would pass the test brilliantly, or he would miss the objective he had set himself, putting the whole credibility of the experimental set-up at risk. Whatever the outcome, it was up to him. Incarnation would be first of all a test of himself.

One of the most delicate interactions he had was with a fisherman's wife who asked him some level-headed questions. Just as Karan felt he had a mission, she too had her own. '*Devījī* [a term used to address a woman with respect], come closer to me. Do you have a question?' he asked her as she shyly approached. 'Yes,' she said. 'I want to know who we're dealing with,' she said, pointing her finger at him. 'What is this thing in front of us?' 'I am the electronic Ganesh,' Karan replied. 'I come from Mount Kailash. I am Ganesha, the son of Parvati. Ask me your question.' The woman reacted strongly, frowning. 'We have never seen anything like this!' Karan, however, felt imbued with a special kind of legitimacy: 'It is the first time that I am on the Earth in this form. My goal is to increase devotion to me, through this medium, this Ganesh ...'

The woman was still sceptical: 'We didn't know that was your aim. Why is there such a need?' she asked. 'I felt I had to come to see you, and to talk to you,' Karan said. 'The Ganesh idols you have seen so far did not speak. But I have the gift of speech. I have come to talk with you and find solutions to your problems. You can find me through devotion and meditation [*sādhana*]. If you see me in this form, it will increase your faith and devotion.' His interlocutor was not convinced and continued to display her doubt: 'Ganesh in his traditional form is fine with us. We don't know what is true and what is false, whether you are a true god or an impostor. We have never seen you in human form.' And Karan replied, 'On this Earth, man must accomplish everything by his own goodness. He alone must judge what is true or false.'

She then subjected Karan to a real test, and by way of him, all the gods. 'If there is truth in devotion, why can't this truth be apprehended?' she asked him. Karan replied that God was the only truth, that she could reach him through devotion and thus fulfil all her wishes. 'But this has never happened to me! I am devout, I perform all the rituals, but the more I do, the more injustice I suffer!' Karan had no answer for her but a promise: 'This injustice will come to an end,' he said. 'Your devotion must be sincere. Worship Ganesha. Worship me.' The woman then asked him if all her wishes would really be fulfilled and if it was possible that her situation could change, just like that, overnight. 'All that you can think of will come true,' Karan promised. 'So far, this hasn't happened,' she replied.

She brewed with frustration and anger, then raised her voice: 'If someone prays sincerely for the fulfilment of their wishes, why do they continue to face difficulties?' Karan replied that her wishes would surely be fulfilled, that the difficulties she faced were the result of her actions, both good and bad. She interrupted him straight away, fiercely: 'Our wishes are fulfilled but we are going through difficult times, we see no way out. Even when we patiently get results, hostile forces take over and make fools of us!'

Serene after a moment's reflection, Karan seemed to have found a way out: 'Life is made up of ups and downs. If your enemies laugh at you, you must ignore them. You must concentrate on your good deeds. Your purpose on Earth should be devotion. Worship God, Shiva, Ganesha ... you must worship them all. Then your wishes will be fulfilled.' And while her bitterness was at its peak, the woman continued to address Karan with respect, if not a hint of irony, 'That's enough! Everything you say must come true. Everyone prays to God. If someone's wish comes true, that would already be something, that would be Heaven on Earth!' And Karan continued to promote a magical god, a god who promises and grants wishes, a god deemed untenable by his interlocutor: 'You must worship me all the time,' he said. 'You don't have to go to the temple. You can worship me at home. You can ask me for anything you want, as long as it is done with conviction. You have to think about your purpose all the time, morning and night. And you have to think about how you can get that thing, all the effort it takes. If you think about it, you will get it, whether it is wealth, recognition, fame or anything else. You can acquire anything in this world, it depends on your attitude.' But the woman was not ready to be hypnotized: 'It would be nice if it were true!' she told him. Despite this, Karan continued in his attempts: 'I will grant your wishes. Close your eyes and formulate your wish but don't express it openly in front of everyone, formulate it in your head.' She closed her eyes for a few seconds, only to open them again straight after.

Curiously, she suddenly turned the conversation to theology: 'Is there any truth in the fact that God exists in our minds?' she asked. Faced with such an unsatisfied metaphysical thirst, Karan could have seized the opportunity to change his method and perhaps invent a novel form of god together with her, in a real dialogue. For that, it would have been necessary, at this very moment, to turn the question around, for Karan to admit that he was just as at a loss as his interlocutor, but he could not, he was too committed to his initial project, to use the machine to plunge people into a modified state of consciousness favourable to the acceptance of all paradoxes. 'God exists in this world,' he told her.

'Brahmaji, the father of the Universe, created man and woman. In this life, the good and the bad are parts of the world. Your goal should be devotion.'

Karan rolled out his ready-made formulas, not able to hear that his interlocutor wanted to embark him on a more profound exercise, a radical questioning of his own being: 'Why can't we see God? The God that we adore so fervently, why doesn't he show himself? Why doesn't he appear in front of us?' If until then she had addressed her recriminations sometimes to God himself and sometimes to the incarnation who was his representative, she had now accepted the ambiguity of the latter on condition that he help her to clarify the mystery of God itself. But Karan provided answers as a machine, ones that were as short as possible, bringing us irresistibly back to the old contract, the one that binds men and gods by the *promise*: 'If we do not see God,' he said to her, 'that does not mean that he does not exist. He exists. He listens to your desires. Ask your question.'

Clearly, he could not see his mission any other way. But she didn't see it that way.

'Why did you come in this strange form?' she asked him again.

Karan's position was not an easy one, admittedly. Unacceptable as a god, always referred to as an intermediary, he had to prove himself as a fuzzy vector. In its tele-operated form, the new variant of the God, a singular mix between a voice of proximity and an image at a distance, allowed repressed reproaches to be said. This was undoubtedly its main interest: to reset all the parameters of God to zero. Karan had to invent something else. Even the invisibility and silence in which the gods had until now willingly taken refuge with the complicity of a whole ritual technocracy orchestrating their presence-absence, were no longer self-evident.

'I am the electronic Ganesh,' Karan said. 'I was created by scientists and by the performance of a ritual, for which I came from Mount Kailash. I have come to converse with you all. Ask your question Deviji ...' 'But I have never seen Mount Kailash,' she exclaimed. 'We are human beings. We only know the Earth!' Karan insisted, 'Without worshipping me, no one gets anything.' The glaring paradox of the situation was that it was he who was now imploring her, not the other way around. When he urged her to think about her wish fervently, she retorted, 'I will ask you to grant it and I will check whether it comes true or not!'

Seeing that he wasn't getting anywhere, Karan made another attempt: 'Tell me Deviji, what are your problems,' which triggered a new wave of anger from her: 'Are our problems hidden from you? If you are God, don't you know our wishes and desires?' Karan tried to be as gentle as ever, 'I know, but I need to hear it from you. And if possible ...' She interrupted him again. 'Is it necessary to say it out loud? If you want to grant our wishes, you must pay attention to our desires. You must know the needs of every human who worships you!' At Karan's insistence, she finally expressed her dearest wish, that her children be educated and successful because 'this is the dream of every parent'. And Karan had this surprising answer: 'Your wish will come true, your children will be educated and they will succeed in doing what they want. For this, you must organize yourself, plan their education. You have to ask them regularly what their plans are for the future and help them when they run into problems.' Against all expectation, Karan had also entered into a logic of subversion. If he had been so keen to establish a magical relationship with his interlocutor, it was to subvert it from the inside and introduce a little pragmatism. 'Will you not give me your blessing now that I have expressed my wish with sincere devotion?' she replied. He then uttered *Oṃ namaḥ śivāya* twice. She did the same and the interaction came to an end with all the respect due to a deity.

Karan was not the only superfluous god in the history of religions. Like an old chat program for computers, in which only a very limited number of possible answers had been taught, Karan always repeated the same messages until he wore out his interlocutor. He could have belonged to the first age of robotic gods, those who use a limited number of formulas and try to plunge their interlocutors into an altered state of consciousness, halfway between hypnosis and meditation, in order to restore their confidence.

His discourse was even more mechanical than his mask suggested. But he was probably one of the few gods to say so clearly that he wanted to attract more and more devotees to him and not just solve people's problems. A kind of narcissistic god, always in need of love and who feeds on the veneration he gets, without ever transforming promises into actions. Before him was someone speaking on behalf of a collective, all those who, like her, performed rituals without ever obtaining benefits truly attributable to the deity. How could Karan have become the defender of this minimal circuit of devotional rewards when his interlocutor brought him the proof of his lack of reciprocity, looking for a way out of it? How could he become the promoter of magical thinking? Wasn't to do so to risk imposture? When one thinks of God, one thinks 'miracle', even if for miracles to happen, he said, there was no need to recite magic formulas, it was necessary that the request be made with 'sincerity' and to think about how to do it well.

Karan was in fact ready to do anything so that his interlocutor would internalize him, so that she would incorporate him in her turn as he had incorporated Ganesh. But in her stream of reflexivity, the woman resisted. It was enough to endow God with a word – or we should say, rather, with an interactivity, even minimal, since he has always had speech – such that all the harshness of the ordinary devotional relationship rises to the surface. The interlocutor doesn't let herself be pushed around. She sows a seed of doubt about his form, his mission, and, through Karan, it is the whole apparatus of devotion as a routine of keeping accounts, the movement of gains and losses, that she puts in question, all the way through to the presupposition of divine omniscience and of the intimate communion of thoughts that constitutes its foundation.

She soon realized that addressing this engineered god was not quite like addressing the god of the temples. She could afford to make more demands and, once the balance of power was established, she could put him to the ultimate test: if ever her wish was granted, she would continue to worship the gods of the temples without a second thought. Karan knew that he would not be called to account and that if he promised something, no one would reproach him for not having fulfilled it. And, confronted with the ordinary preoccupation expressed by his interlocutor about the future of her children, Karan played his role. Here as elsewhere, he took advantage of his position to hand out techniques for positive thinking. Who could blame him? He had wanted to be a good incarnation, which meant, for him, to distribute hope by any means. His interlocutor was not desperate, but it did not take much to stir up her doubts, which had previously been deeply buried. Initially, she hadn't had any particular question to ask. It was He who became the question.

Regarding each incarnation's capacity to arouse doubt, there was a learning curve. These doubts were mostly expressed with a certain respect. The more people Karan interacted with, the more he learned, gradually shifting his ready-made answers. 'I reside in the idol. I go to all the places where my idols are placed. But this time, I appear though an electronic medium. I can talk to you through the interface. I can solve your problems,' he said. A famous idol maker approached the machine. 'I have no doubt about Ganesh', he said, 'my doubt is only about you. We give such importance to Ganesha that we cannot desecrate him. We respect Ganesha as he is. We might be less careful about the goddesses, but we are very careful about anything that might introduce movement into the idol of Ganesh. Because if we do that, problems may arise. That's why we leave them [the idols] still. So yes, there are questions. Everyone here is wondering about your electronic form. Is this new form really the idol of Ganesha? You may be a marvel of science, but you are generating some confusion.'

Karan had not suspected that he would arouse so much suspicion. By mimicking meditation, he had unknowingly kept God as close as possible to the stillness in which idol makers like to fix him. The absence of movement avoided any question-and-answer game. To justify his ability to move and speak, the incarnation didn't have that many options. He could either bet on the total symbiosis between God and himself, or he could argue that he was only a representation, an actor wearing a mask, but this was just as unacceptable. Thus Bappa gradually individuated himself, from one embodiment to another, first of all taking for granted the fusion between God and the incarnation, whereas this was precisely what was at stake, and far from guaranteed. Very quickly, it became apparent that the supposed fusion from the outset was not viable. One had to learn to be God before proclaiming oneself as God. Karan learned this the hard way. Yet he was just beginning to improve his conversational curls: 'I am the same Ganesha. I have two ears. I have a trunk. You can touch it and feel it. I can hear and feel your presence in my trunk. You can touch my trunk. I travel from place to place with the rat as my mount. Don't be shy. I have come here for you. I have come from far away, from Mount Kailash, just for my devotees. If you do not come to me, I will disappear forever.'

3

Kaliyuga

Whilst the robot had barely started its tour through Mumbai, we were inundated with requests. Families wanted us to bring the robot to their homes. The officiants would have no problem with it being present, quite the contrary. The Mahindra family, whose father was a banker, had a *paṇḍit* brought in for a ritual. The robot was placed next to the altar where the idol of the house was located. The *paṇḍit* put on the helmet and began his invocations. The invocations were transmitted via Bappa, which gave the face of the officiant an almost comical image. No one saw this distortion as a problem. Bappa was still just getting into its stride. We were trying to better understand its range of applications and whether it could be of any use, whether it could be integrated into the ritual or not. If he had been endowed with his own programming abilities, like a learning robot (which it was not), this new situation would surely have been a good way for him to get into his role, by modelling himself on the words and actions of an officiant in full action. But Bappa had no memory of his own other than that of his operator, and he would not be fulfilling any ritual role other than that of amplifying the officiant's words. When one knows the importance of the formulas pronounced in a ritual and the need for them to be heard by the god being invoked, the amplifier function was not negligible.

In any case, the *paṇḍit* was too busy with what he was doing to worry about his enlarged image or the echo that now enveloped his speech. A thousand leaves of *dūrvā* (Ganesha's favourite herb) had to be piled one after the other around the idol to promote the growth of the host's family lineage. The ritual lasted for several hours and the family members took turns in this demanding task. According to the *Gaṇeśa Purāṇa*, Ganesha was to be worshipped with this herb on the first day of the Ganesh festival. Misfortunes and obstacles would disappear. The ritual performed among the Mahindras was meant to be an exact replica of the ritual performed by Vaishravana, a king celebrated in mythology for being an exemplary devotee of Ganesha and having obtained many things.

The officiant mentioned the eleven types of divine desires that the god was to fulfil: Devotion, Honour, Success, Wisdom, Knowledge, Intelligence, Happiness, Strength, Long Life, Glory and Health. And the god was to give his blessing to all their guests and friends who consumed offerings that had been offered to him. If any of them had blemishes in their horoscope, the god was to remove them through a ritual of invoking the sovereign principles (*Āditya*) or appeasing the nine astral houses (*Navagraha śānti*). The officiant asked that these rituals be performed with all the necessary faith and devotion and that everyone be inhabited at that moment by good thoughts. The gains were to be irrevocable and assured. He invoked the eight forms of Goddess Lakshmi, *Dhanalakshmi* (goddess of Wealth), *Dhanyalakshmi* (goddess of Grain and Food), *Bhagyalakshmi* (goddess of Luck), *Saubhagyalakshmi* (goddess of Fortune and Marriage), *Santanalakshmi* (goddess of Fertility), *Tarunyalakshmi* (goddess of Youth), *Shauryalakshmi* (goddess of Strength) and *Dhairylakshmi* (goddess of Courage). The wishes and desires of each devotee were to be fulfilled as soon as possible.

Then the *paṇḍit* asked the host couple to make three salutations followed by three rotations and to perform an eight-pointed floor salutation posture (*Sashtanga namaskār*), before sitting down to recite mantras. He carefully guided the couple, asking them to repeat each word:

> Now you must pronounce the word *Runā* (obligation), *Rogādi* (sickness and misery), *Dāridrya* (poverty), *Pāp* (bad deeds), *Apasmāra* (epilepsy), *Mṛtyavaha* (death), *Bhayā shoka* (fear, sadness), *Manastapaha* (mental anguish), *Nashyantu* (destruction), *Mā* (mother), *Sarvadā* (eternally), *Śānti Astu* (may there be peace), *Śivam* (favourable), *Astu* (let it be), *Graha* (house), *Kuruvantu maṅgalam* (favourable).

The host then had to throw grains of rice (*akṣata*) at the god, like confetti. Then another sequence of invocations:

> Created by illusion (*Māyākṛta*), we bathe your head (*Sirabhiṣeka*), O Lord of categories (*Gaṇapati*), we perform the invocation of auxiliary mantras (*Aṅgapūjā*) for you, the awakening of the circuit of the goddesses (*Āvaraṇapūjā*), worship by leaves (*Patrapūjā*), worship by flowers (*Púṣpapūjā*), accompanied by (*sahita*), hundred (*sāhastra*), names (*nāma*), through (*dvāra*) the work (*Kṛta*) of the Durva leaves (*Dūrvāṅkura*), which results (*Arpaṇa*), by the action of the herbs (*Kārmaṇa*), on the family (*Saparivāra*), the earth (*Prithavi*), Ganesha master of powers (*Siddhivināyaka*), heavenly beings (*Devataha*), the one we cherish (*priyata*), He who has no equal (*Anenā*), the Supreme Being (*śrī Parameśvaraha*), that which is mine (*mama*), Divinity of the family (*Kuladevatacha*), our beloved (*Priyatama*).

Then the *paṇḍit* asked the host to pour water and throw rice on the idol again. He told him to hold his hand properly and repeat after him:

> Please give me Power (*Shakti*), Well-Being (*Swasti*), Existence (*Astitvam*), Faith (*Shradham*), the Vedas (*Vedam*), Success (*Yashaha*), Intelligence (*Pradnyam*), Knowledge (*Vidyam*), Fearlessness (*Abhyam*), Opulence (*śriyam*), Strength (*Balam*), Long Life (*Ayushama*), Brilliance (*Tejas*), Health (*Ārogyam*), O Ganesha master of those who attend to the gods (*Gaṇanāyaka*).

The officiant then asked the host to take the flowers, smell them, pour water on them and repeat the words:

> Work of Illusion (*Mayakrita*), work of the rite (*Pūjā*), let him join (*Sangata*), accomplish (*Siddhyartam*), to him (*Asmai*), to the master (*Ācāryaya*), in his power (*Yathāśakti*), the fruits (*Phāla*), through the offerings (*Dakṣiṇam*), for you (*Tubhyam*), me (*Āham*) and our lineage (*Sampradādeha mama*).

'Now you must pour the water,' the officiant said. The host poured the water and put back a layer of rice grains. The officiant made the offering of *Pañchāmṛta* (a mixture made of milk, yoghurt, honey, clarified butter and banana). Another round of mantras and chants followed by the consumption of the offerings. The host thanked the *paṇḍit*, saying, 'This ritual was very pleasant. I feel very happy.'

During the whole of the ritual, the poorly fitting helmet on the officiant's head kept slipping off, making his tele-operated face grimace. We should have been careful with all these little adjustments. We too were getting through the teething phase. It didn't take much for the whole thing to slide into parody and to be experienced as an attempt to denigrate the ritual institution through the intrusion of an exogenous accessory. And yet, why was such a misunderstanding, in actual fact, difficult to imagine? No one, apart from the designers of the device, seemed embarrassed by the ridiculousness of the situation. It was therefore up to us to do a self-analysis.

In order not to find the tele-operated image grotesque, we had to break with a well-established habit of perception which holds that poor transmission spoils the source image or that the deformation of a face can only denigrate the source. Perhaps the relationship between the original and the substitute also had to be viewed differently, in favour of the latter. Brahmanic speculations on this subject are interesting. They regulate the use of substitutes for many rituals; because the original is unique it is always absent by definition, or is untraceable, whereas the substitutes are plural. The *dūrvā* herb used here, for example, is a substitute for *soma* (the nectar of the gods). But it is a fourth-order substitute which the texts tell us is advocated only if one does not have *phālguṇa* plant available (second-order substitute) or *ādhāra* plant (third-order substitute). And if *dūrvā* is not available, a multitude of other substitutes are possible (the *kuśā* plant, for example). In practice, rituals are made of cascading substitutes, the *paṇḍit* operating with second- or third-degree substitutes, i.e. substitutes of substitutes of substitutes ... Certainly, Bappa did not claim to substitute in any way for the officiant. His ritual function remained undecided and ambiguous. And we must acknowledge the open-mindedness of the *paṇḍit*, who willingly lent himself to the game, to see what the use of technology might change. It is important to note that the question is what such recourse to technology would bring about for the ritual and not for the officiant himself. We would be missing out on a whole host of potential uses if we only saw it as a means of enhancing the person, or the image of the incarnation.

Indeed, using the interface as a means of augmenting the officiant, making him Ganesh, made as little sense as trying to force the god's entry into the ritual. When the entity aimed at can be summoned only by gestures and formulas, there's no way to make it turn up automatically, and it is never certain that the entity in question will respond. If we are willing to overlook the risk of grotesque disfigurement, Bappa appeared here as a simple prosthetic detour, hooked up to the non-human realm (more than to the human) in a ritual machinery whose high degree of technicality the reader will certainly have noticed, before we even thought of introducing an electronic artefact.

If he had been concerned about his image, the *paṇḍit* would no doubt have pointed out that these design flaws in the helmet should have been corrected. And if the masters of the house had realized that Bappa could turn a ritual into a farce, they would probably have said as much, and would have asked us to stop. It's that in this case it was not a question of *incarnation*, only of transmission. In a face-to-face interaction, God is embodied (or not). But in a ritual, the challenge is not the same, the divinity must be summoned with the maximum intensity, all its senses as well as our own must be solicited, power must be transmitted via words, substances, gestures and, if possible, with the minimum amount of leakage or loss in the connections and flows. Through his insouciance, the *paṇḍit* had unwittingly reduced Bappa to a pure amplifier of information in the thermodynamics of the rite. What we needed to test was Bappa's capacity to serve as an energy relay, even if no one could really say at this stage whether this addition to a well-functioning ritual machinery had been successful; only the family sponsoring the ritual could, one day.

By ignoring the distortion of the image, not only had the powers of the grotesque been neutralized, but a whole new field of experimentation was opened up, by and with the substitute. We had to grasp everything that could potentially cling to it, resonate with it, be amplified by means of it. Can we imagine altering a signal, a vibration, in such a way? Can a mantra be transmitted electronically without loss? Does an echo in a recitation make it more effective? To what extent can we strengthen the rite with all sorts of technological amplifications and derivations? We are not the first to ask these questions: those artisans of ritual platforms who shamelessly tinker with the gods – and who do not see this as an attack on divinity but, rather, as a means of achieving it – have been asking them for a long time. Every time these platforms use new special effects, strobe lights and other psychedelic effects such as those I described in *Gods and Robots*,[1] the transmission properties of the medium, its conductive potential must be assessed. And what if Bappa had only been an emblem for the *paṇḍit*, the admirable but imperfect representative of Technology, a bit of modernity offered as it is to the god, the better to arouse his curiosity or even to be in tune with the times?

Once the ritual was complete, the *paṇḍit* took advantage of his amplification, via Bappa, in order to share his thoughts, hoping that they would be relayed, somewhere, perhaps to a wider audience or on television. 'I have to tell you some things,' he began. He had plenty of stories to tell of well-performed rituals that produced a boon. 'Each god reacts to certain things. For example, the god Surya (*Sūryanārāyaṇa*) is pleased when one bows before him. Similarly, one should offer *kumkum* (red powder) and flowers to a goddess to please her. The god Shankar is appeased by making an *Abhiṣeka* (milk offering), while Ganesha likes sweets (*modak*). This makes the god happy.' While the god does respond when invoked, he remains nevertheless unmoved by any nonchalant invocation. Intensity of devotion is what he appreciates.

'On the fourth day of the *Gaṇeśa Chaturthi*', the *paṇḍit* recounts, 'in the lunar fortnight of *Bhādrapada* (entering the twenty-sixth constellation), devotees of Ganesh chanted the seed mantra (*Bijakṣara mantra*) and attained enlightenment, having recited 2,800,000 mantras that contained tantric words and prayers to remove obstacles. Ganesha's rat was passing by. He asked him to visit the houses of the devotees. But as this would harm his feet, he asked for something soft like *dūrvā* grass. From then on, whoever worshipped with *dūrvā* grass would have his wishes fulfilled.'

The *paṇḍit* then multiplied the 'true stories' demonstrating that the amount of mantra and the intensity of devotion unleashed a power to which the capricious god never remained insensitive. 'One day an ascetic visited a couple who had no children, and who had been given little chance by doctors of ever having any, for various reasons. The ascetic told them that if they performed the right invocations to Ganesha along with the *dūrvā* offerings, they would eventually get a son who would look like Ganesha. Three months passed. The couple performed many rituals. After a year, they were successful. The child had big ears, a pointed nose and slightly crooked teeth. He looked like an elephant. Later, they built a temple in Ganapati. Their child was the only son in the family of four brothers.' This was a clear sign to the pujari that there was a link between the intensity of devotion and the achievement of a result.

The *paṇḍit* had more than just rumours or mythological stories of exemplary devotees. To highlight Ganesha's effectiveness in the search for prosperity and in establishing virtuous circles, he wanted to share his own experience:

> A man had a court case that had been dragging on in the Supreme Court for forty-two years. The case had been transferred by the High Court. I advised the person to do the well-known ritual of paying homage to Ganesha (known as *Arthavaśīrṣa*) and to place one thousand *dūrvā* leaves on Ganesha every month at the appropriate time. Thus, the person performed four thousand *Arthavaśīrṣa* rituals for four months, supplemented by an *Abhiṣeka* [pouring of milk on the idol's head]. A few days later, the Supreme Court judgment rendered. Proof again that we get real results when we invoke Ganapati: But some people do not believe in this. They do the ritual minimally and expect maximum benefit. Nobody gets anything that way, because it is not the right method. Mantras and Ganesha rituals have a tremendous effect of which we are hardly aware. What we know today is just a drop in the ocean. We perform the rituals by tradition, without ever questioning. But we should first understand the power and purpose of mantras. We have to understand the meaning, perform the rituals according to the correct procedure and do them regularly in the right way. Now, what happens in the Age of Destruction [*Kaliyuga*], the moment in which we find ourselves? All mantras were created during the *Dwaparyuga* and *Tretayuga* [the second and third ages according to Hindu mythology]. At that time, the meaning of mantras was known. During the *Dwaparyuga*, all these mantras only needed to be chanted once for good results to be obtained. It was enough to make a single salutation to the god. But in *Kaliyuga*, we have to do four salutations and we have to do many more rituals and mantras. We don't get the same results as in *Dwaparyuga*. So we have to do more rituals to get similar results in *Kaliyuga*. Before, one mantra was enough, but now we have to include as many mantras as possible in the rituals. And these must be done in the right order and systematically. One should not imagine that all this can be done in one day. Each mantra may take four, five or eight years to be fully understood, heard and to generate results. Without a guide who is experienced in this field, one cannot harness the power of mantras and their benefits in real life.

The *paṇḍit* wanted to remove the helmet from his head. In the end, it had become uncomfortable. He had properly defended his vocation, calling for more and more rituals, more mantras, more specialists to revitalize them and more research to prove their effectiveness. At some point, it was inevitable that the *paṇḍits* too would come to lament via Bappa, to make the most of it so as to convey their concerns, or their truth. His appeal was not only to the devotees, but also to his colleagues. It is hard to imagine a *paṇḍit* saying that his whole body of knowledge, his substances, his formulas and his frighteningly precise gestures have only a psychic effectiveness, only a simple effect of 'positive thinking' on those who follow them scrupulously, even if it happens that *paṇḍits* often do hold such discourse. It is equally difficult to imagine him defending a stupidly mechanical conception of effectiveness, as if the effects of the rite were guaranteed. According to him, their effectiveness was real, their resonances ignored and their signification lost, for the most part. And thus, if it was necessary not only for the ritual acts to be well done, but also for the devotion to be real and intense in order to awaken the god's sensibility, failure to get a result was necessarily the product of a defect in form, of a poorly performed procedure, of a mantra poorly uttered, a vague gesture or of a lack of fervour.

I couldn't help but think that next time we really would have to make sure that the helmet fitted properly, to show the same care for our wires and connections that the *paṇḍit* showed for his own toolkit, or else we would risk undermining the process. As I listened to his last words, I better understood the first officiant's attitude when performing the ritual of opening the eyes on the machine. Against all expectations, and without wishing to generalize to all *paṇḍits*, it was a very unique way of experimenting in metaphysics.

Metaphysics is usually considered to be a matter of thinking rather than doing. Let's venture another definition. Any activity is potentially metaphysical from the moment we grasp its intimate or hidden resonances, its repercussions, even the most counter-intuitive, at levels other than that of immediate perception. One has to be willing to let oneself be led astray, to the point of vertigo, through the lists of formulas and ingredients of Hindu rituals in order to grasp that it is an experience and experiment in *concrete* metaphysics that is a matter of doing, or more precisely, of the order of a coalescence (*concrétion*) in act: physical operations with *meta* implications are experienced, via gestures, things, substances. And there is certainly a close link here between the exuberance of rules and operations and the invisibility of the entity addressed.

The more the thing is in effect absent, the more we act to make it exist. It would not be a distortion of the disorder of Brahmanic speculation to assert that it is traversed by certain principles that still animate the most banal rituals today, such as the one we witnessed, and that the idea of a *driving absence* is at its core. One could invoke other equally essential principles, such as that it is incompleteness that activates, that it is the only creative power, that completion is considered unattainable and dangerous, that only the false, only error and lack are generative, or else that endlessly ruminating on the indefinite is better than seeking the infinite. But I prefer to move forward gradually, to stick for the moment to what we can experience via Bappa and to let it reveal its own lessons.

Making an experimental god from scratch, as is by now clear, did not make much sense here, since it is only possible to experiment with substitutes for a form that is always absent. On the other hand, within the field of substitutions, there is in principle no limit, and it is entirely possible to reset the properties of the absent entity to zero on the basis of its substitute, to modify its attributes, to test others, and so on. If we want to experiment with the properties of a new medium, nothing is impossible, on the contrary, any new technology is welcome, as the *paṇḍit* showed us. It is necessary, however, to accept that metaphysics and technique (or even bricolage) become synonymous to the point of being entirely confused in the experiments with (as well as the experiences of) powers and effects.

In the age of *kaliyuga*, as the *paṇḍit* reminds us, it is not only men who must rethink their ways of harnessing divine energies to their advantage. Everyone is put on a learning curve, including the *paṇḍits*, who must relearn how to handle the power of the formulas and adapt them, for what we know of their possibilities is but 'a drop in the ocean'. One can be content to follow tradition in this field, but it is entirely possible to conduct one's own experiments; just as Colonel Albert De Rochas had fun experimenting with his neighbour with multiple forms of action at distance in order to question the foundations of the practice of bewitchment and invented, almost by accident, bewitchment via photographic plates (*L'extériorisation de la sensibilité*, 1895). Here again, the *paṇḍit* guides us: mantras must be experimented with. We are far from having finished (re)discovering or investigating the vibratory properties of sound: 'Some attribute the effectiveness of mantras to their vibrations, others to their power of psychic stimulation. The therapeutic benefits of mantras are listed in Ayurvedic medical treatises and every day new experiments demonstrate their virtues.' Even when indexed to the procurement of a gain as elusive, as laborious to obtain and as difficult to measure as prosperity, it was clear to the *paṇḍit* that only by learning the lessons of the past while experimenting with new virtuous circles combining technology and invisible powers, would men maximize their chances of success. And while there was no doubt in his mind that, among the gods, 'Ganesh is the most bankable god', it was up to each individual to experiment, to test this power for himself and to judge the effects.

'Try Ganesha and you will see,' he said.

Note

1. Emmanuel Grimaud, *Dieux et robots* (Apt: L'Archange Minotaure, 2008).

4

Divine (In)Justice

Karan had been really getting into the game of incarnation, until the day someone tried to catch him out. He was a bus conductor named Nitin Parwar who was also the head of a *mandal*, one of those neighbourhood groups that raises funds during the Ganesh festival, makes an idol for local people and organizes the appropriate rituals. The organization had set up a scaffold in the middle of a small square with an idol three metres tall, which visitors came to worship. Nitin was also known by his neighbours for his involvement in charity work. He agreed to have our robot set up nearby with a few chairs, which would allow anyone to play the incarnation game and watch the show. And to set an example, Nitin Pawar was the first to strike up a conversation with Bappa, surrounded by a bunch of children from nearby schools, all sitting quietly.

'As you know, Mumbai has a big traffic problem. The city has become very congested. What solution do you suggest to solve the problem?' asked Nitin Pawar. 'This is a human problem. The solution lies with you,' said Karan apathetically.

'We are trying solutions', Nitin continued, 'but I want to know what yours is. There are many accidents. Young people don't follow the rules. They are enthralled by speeding. What advice would you give them?'

'I would advise the young people to slow down,' stammered Karan.

Nitin was surprised at such a trivial answer: 'That's a trivial answer ... what about potholes on the roads? The municipality does not take any initiative, even when we write letters to them. What message would you send to the corrupt officials?'

Karan sank into silence, and after a long pause, let out a 'no'.

'No, what?' Nitin asked.

'I am thinking about the problem,' the incarnate replied.

'Oh ... you're thinking ... As you are Knowledge Incarnate, you should have answers ready. Is my question so difficult that even Ganapati Bappa has to think about it?'

Karan looked a little unsettled. 'No, my son. I am thinking about something.'

With each silence, Nitin pursued with ever more vigour. 'Okay, so even God has to think when it comes to corruption? Bappa, everyone's waiting. You seem to be more and more immersed in your thoughts!'

'These things should be avoided as much as possible,' Karan countered.

'What should we avoid?' asked Nitin.

'We should neither take nor give bribes,' said Karan.

'But it's a trend! Even to get a place in a school, you have to give a bribe. How can this problem be solved?' insisted Nitin.

This time Karan's response was immediate: 'This problem was created by you. The solution lies within you.' As before, it seemed that Karan knew a secret formula for arousing the anger of his interlocutor.

'You provide no constructive solutions,' exclaimed Nitin. 'You shirk all responsibility. You say that these are man's problems. But God himself created the mind (*l'esprit*) of man. Man is not born with a computer for a mind. This mind itself is responsible for all these problems. So how then to clean up men's brains? [*Silence* ...] Bappa ... are you thinking again?'

Nitin's digs systematically drew a laugh from the audience. I did wonder, though, if Nitin was as aware of the device as we were. How could he disregard the machine, deliberately discount the subtle mechanism of tele-presence, in order to focus on the man facing him? The purpose of all technology is to be forgotten. In long-distance communication, we are even able to ignore its unbearable cost, in terms of energy, as well as the cables that riddle the oceans, the data centres, the monstrous paths that our signals follow on a global scale. Nitin seemed to show the same feigned ignorance. But, as we will see later;

it was all strategic, at no point had he forgotten the device. He simply wanted to play the game and get straight to the point. A confrontation was to take place between him and God. Or, rather, between God and us, because he had managed to take the crowd with him, in whose midst we were sitting and whose representative he was. Whereas previously the medium had been the object of an array of interrogations, functioning as an impediment to communication, it had now become peripheral. No one paid any attention to the way in which speech was transmitted through the machine, and everyone pretended to take the god seriously, submitting to him a whole host of possible and unimaginable problems. The incarnation was once again trapped, but in an exactly symmetrical way: not because it was doubted that he was a god, but because there was a wish to show that the god was *only* that.

After thinking about it, Karan started again:

'You should not give bribes.'

'Without bribes', Nitin continued, 'even the dead can't get cremated. To have the ashes collected after the last rites, one has to pay. This is my personal experience. It is not man's problem. You are blaming man. So what should be done then to purge his brain? These people should be shot or publicly slaughtered!'

Karan sensed that a trap was being set, and his silences grew longer and longer. 'No,' he said. 'We have to work together to find a solution.' Nitin pretended to accept him as a god in order to better put him to the test. Karan, however, could not come up with a satisfactory retort.

'What is your solution?' Nitin asked. 'That's what I want to know. You are Knowledge Incarnate!'

'Do not give or accept bribes,' Karan repeated.

'I say again', Nitin continued, 'that bribery has become a tradition. Corruption is everywhere. The common man has to face corruption from birth to death. Even after a birth, nurses, doctors, all have to be greased. We have to give bribes even for a death certificate. I understand that all these problems are man-made. But you must have thought of a punishment for these people. How to punish them? In my opinion, these people should be publicly shot.'

Karan did not seem to agree with this solution. 'No, not at all,' he replied confidently, but Nitin insisted:

'Then what is the solution? Bappa, what are you thinking? Are my questions muddled?'

Karan was still more silent, lost in thought and Nitin was growing ever more impatient:

'You can keep thinking', he said, 'but everyone is watching. Everyone wants to know why Bappa thinks so much! Bappa is the god of Knowledge! Everyone wants immediate answers from Bappa. Does Bappa need to do his homework?'

At that moment, Karan had only one desire, to escape, to dissolve, to return into the state of stone, to take refuge in deep meditation, the only viable state for the god he wanted to be. And he had the bright idea of sketching out a retreat by clinging to the only formula that came to his mind at that moment: '*Oṃ namaḥ śivāya* ... *Oṃ namaḥ śivāya* ... *Oṃ namaḥ śivāya* ...'

But Nitin was not impressed at all, 'Singing *Oṃ namaḥ śivāya* is not the answer to my question. Anyone can chant *Oṃ namaḥ śivāya* and run away from his responsibility. *Oṃ namaḥ śivāya* cannot be an answer, sir!' he exclaimed.

'I was chanting the mantras and also thinking ...,' Karan apologized.

'Chanting mantras? Bappa is thinking and everyone is looking at Bappa. Everyone wants to know. The expressions on your face have changed. Bappa is irritated now. Maybe Bappa didn't do his homework?' Nitin looked over at the crowd, whose laughter was becoming more and more mocking.

Ganesh's silence spoke volumes. Someone more experienced would have done better, but the fact that he took refuge in silence at a time when he was expected to publicly demonstrate his wisdom was significant. It is enough to endow a god with a faculty of speech and for a human to assume the right to speak on his behalf to make a previously invisible variable appear. *It's better for the gods to remain silent in order to remain gods.* Karan took his time, but he pulled himself together. 'Okay, tell me what situations you have had to give bribes?' Nitin hastened to list them: 'In hospitals. It's not funny. Listen to me, this is my personal experience. When my father was admitted to the hospital, I had to pay to carry his body to the cremation. And I am not the only one. There are other people here too who have experienced similar incidents. But nobody has the courage to say something.'

Karan shifted for a moment into the position of an empathetic inquirer, a posture that he would not hold for long: 'Have you filed a complaint with the police about this?' he asked.

'Yes, we did file a complaint. But no action was taken. They tell us that corruption is a fact of life. They say they can't do anything.

Eighty per cent of the police are corrupt and demand bribes. All the people sitting here have experienced such things themselves. There are many children here who had to give INR 150,000 in the guise of a donation so as to be accepted into a school. They stand in a queue for a whole day to get a chance to be admitted and they are forced to donate.'

Karan then became the representative of the established order and this was the beginning of a dialogue of the deaf:

'The money you give to get into a school is in order to get quality services. The more money you give, the more services you get,' Karan said in a mechanical tone.

'So what can a poor person do? Where does a poor man's child go?' Nitin asked, to which Karan retorted, 'Go to the public schools.'

Nitin was not convinced. 'We have to give bribes for the public schools as well,' he said.

'Aren't there entrance exams open to all?' remarked Karan.

'The costs are not covered. Money is charged for other things, in the name of charity, but that's corruption!' Nitin declaimed.

Karan told him that this was not true, but Nitin was sure that it was, and was sorry that God was so ignorant of the situation:

'If you want to be admitted into higher education you have to pass entrance exams,' Karan said. 'Those who pass the exams are admitted.'

For Nitin, there was no doubt that Karan was completely out of touch with reality, speaking of another era when admissions were made on the basis of merit. 'Bappa, we are not going to put an end to corruption. Right now, Mumbai has become the most corrupt place, including the police and politicians. So, I think you don't have an answer to this problem.'

Karan would have liked to roll out the ready-made scenario that he had carefully developed, a hybrid of mythology and science fiction, but the poor guy found himself faced instead with problems of corruption and school entrance exam fees, mired in the very real issues of local politics. For Nitin, these were probably the only questions worth asking, so as to find out of what this pretender to godhood was really the embodiment.

And Karan couldn't do better than to stir up his interlocutor's rancour. 'This problem was created by you,' he said.

'You give me the same answer. You don't suggest a solution,' Nitin said.

'But you yourself have the solution,' Karan retorted.

'Yes, I have the solution for that. I will shoot them if I have permission! It is the only solution I have. Every corrupt person must be imprisoned and killed!' Nitin proclaimed furiously.

'This problem must be eradicated from our homes for good,' Karan went on to say. 'When you have a small child at home, you corrupt him by giving him chocolates. Isn't giving chocolates to a child as bait a form of corruption? Isn't giving a reward to a child who has passed his exams akin to a bribe?' Nitin agreed. Under certain conditions, giving sweets to a child could be considered a form of corruption or bribery.

'So the practice of bribery was initiated by you,' Karan concluded. Nitin's irritation was at its peak: 'You are repeating the same song again, absolving yourself of any responsibility. You keep harping on, saying that you have not done anything ... As you are the Giver of Knowledge, we just want to know. Please instil some manners in these corrupt people. Some parents sell their jewellery to pay for their children's donation, or mortgage their houses. If this situation was created by humans, then humans should be exterminated. Why does Ganesha have a sword [*parśu*] in his hands? Can you tell me why, since you have taken the form of Ganapati?'

We sat helplessly watching this unusually violent debate. Nitin was not the only one who took advantage of his encounter with Bappa to vent his difficulties and turn the set-up into a court of law. Was there another way to do politics? Seeing Karan struggling, perhaps it should be deduced that it is necessary to change the incarnant, that having Ganesh in you is not enough in order to have an answer to everything. But it was, in the end, Nitin who indicated a possible way, so that the experiment would not be a long, uncontrolled skid, oscillating according to the quality of the participants. Neither unexpected events nor failures would put an end to the process; quite the contrary. The more powerless the incarnant was in appropriating transcendentals, the more Nitin did everything in his power to take the god by the hand and teach him to be a god. The best way to avoid falling back into the old politics of claims and promises was to consider this god as an apprentice, and to teach him his role.

'Ganapati Bappa is armed with a sword so as to eradicate evil impulses,' Nitin said. 'If Ganesha is not able to come to this Earth, to punish these corrupt people, we ask him to pass on his powers to us so that we can inflict proper treatment on these people!'

But Karan was not prepared to go that far. 'The power to change things is within you,' he said.

'Not at all,' replied Nitin. 'We have to give bribes, even to vote. Today, 80 per cent of the politicians feed themselves on bribes and false promises.'

'But aren't you the ones who elected them?' asked Karan.

'You're putting the blame on others again,' replied Nitin, with irritation. 'Please don't put the burden on others. Be constructive! If humans are responsible for everything, then what is God doing?' Karan's answer was clear and unambiguous: 'God does nothing,' he said. 'God doesn't do anything?' Nitin said back to him. 'Well, God did create this universe,' Karan clarified.

'And all he asks of man', Nitin added, with a hint of irony, 'is to live as he pleases? God created the universe and gave man the freedom to live as he wishes? Certainly not! Birth and death are in God's hands, as are intelligence and the power to think. If all this is not in the hands of God, then humans must reserve the right to punish corrupt people. And if God chooses me to do this task, then the only punishment I will give them is death!'

By a curious effect of dispossession, Nitin did not feel entitled to act in his own name, as if God had confiscated all permission to act and the divine firewall had to be lifted urgently in order to regain his abilities. Nitin was desperately seeking the support of a *fuzzy power* aiming at worldly matters in general, not directed at any particular object but potentially interfering in all earthly affairs. It was precisely this indeterminate energy that Nitin sought in order to redirect, that he had to seduce, and with whom he sought to re-establish an alliance. But this is not all that should be seen in this dialogue, otherwise we would be missing the substance.

There was a crisis in the contract between Nitin and his god, there was something twisted, eroded, or even outdated about it, which he sought to put to the test with irony and even a touch of perversity, by rediscovering the purity of a power struggle with God. In order to resolve the crisis, his tactic consists in posing, consists in imposing, an insoluble moral problem on the incarnant. This is a test of contortion, from which Ganesh, the people's friend, this moral and responsible god, must emerge twisted, disfigured, even monstrous.

The object of misunderstanding between them is not so much the existence of God, it is, rather, a technical problem, a problem of *circuitousness*[1] in Alfred Gell's sense of the term. Karan becomes the unfortunate guinea pig in an experiment in limit-experiences that aims to test the extent to which a devotee can trap a god and reverse the course of the transfer of power. How do you circumvent the ventriloquism? How do you bend God, wring him until he is disfigured? How to distort morality? The dilemma of the incarnant is understandable, preferring to remain in the background. Moreover, he continued to withdraw, ever more powerless, and then began to invoke other figures.

'The good and bad deeds of corrupt people are counted by *Chitragupta* [the god who keeps the record of men's deeds],' says Karan, in a tone just as machine-like as before.

'So make me *Chitragupta*, let me be *Chitragupta*! Or make me *Yamadut* [a messenger of Yama, God of death],' beseeched Nitin.

'They will all be punished by *Chitragupta*,' Karan said.

'But *Chitragupta* cannot come down to Earth,' Nitin replied.

'*Chitragupta* records all the details of the actions of corrupt people', Karan clarified, 'like a scratched old mythological record.'

Nitin retorted with renewed vigour: 'We know that *Chitragupta* keeps the records of deeds, but who has actually seen him? Up until death, no one sees him come down to Earth, and after death he doesn't show up either. So, how could we know what punishment *Chitragupta* inflicts on the corrupt?'

Karan could have been happy, he and Nitin were at least using the same images. It is amusing to see how deftly Nitin played with the device. He never put in question the idea that God could be incarnated in the form of a person, and with a certain irony he used this ease of communication to push the idea to its limit, as if the game of incarnation was the perfect opportunity to replay everything, to rewire worn-out divine circuits, to subvert fusty personifications, to submit mythological images to the test of reality. How to recover the field of powers? Reconnect with that of which the gods are but the name, yet which has been lost by assimilating them to people? Karan had no other solution than to make his own mythological film and to bring Nitin on board. Nitin, for his part, had no choice but to try to subvert this film from within by all means:

'Everyone will suffer the consequences of their actions when the time comes,' warned Karan. 'We are in the *Kaliyuga*. We are no longer in the age of *Satyayuga* where I could act with a little magic and appear to you on command.'

Did Karan know at that moment where this dialogue would lead him? Nitin was not only angry at the corrupt, not only angry with the politicians, he was also angry at the gurus who claim to 'do magic and perform miracles':

'Who are these people then [gurus]? Why do people admire them so much? They are only conjurers, not real ascetics, they have no extraordinary powers. They do magic tricks! They do magic tricks to seduce you. They take advantage of your innocence. They drive luxury cars and you queue up to see them.'

And Karan simply drove the point home. 'And all because of your backing, the people's support. Why prop them up?' he said.

'Just as we have to get rid of the corrupt, we have to fight the *babas* and de-indoctrinate the idiots who are being taken in by them!' exclaimed Nitin, scarlet with anger. 'You'll probably get rid of, won't you?' asked Karan.

'If given the chance, I will take a hockey stick and hit every one of them. I'll strike them all down!'

Silence.

Nitin calmed down a bit and took the hand of a little girl who was standing to his right.

'This little girl wants to ask you a question too ...' he said.

Karan and Nitin had braved each other to the point of exhaustion. They needed a break. Confronted with such a singular challenge, divine speech had never been so hard to find. It must be grasped just what is at stake in the very moment when the incarnant gets stuck: he fails to appease his interlocutor. Perhaps it is in those moments of bewilderment, of hesitant confusion, when Karan pauses and when nothing seems to be happening, that the emptiness is felt as pure driving power. This is a reading that can only be hinted at, at this stage; we will have the opportunity to return to it at length later on. The dialogue with Nitin would not have had the same substance, the same dynamics, without the moments of taking a breath, those slips and stutters, unknowingly enacted by the incarnant. And this is undoubtedly one of the ways of generating a bit of god, *through the void*, toward which Karan unwittingly pointed.

Given how deeply uncomfortable it was to embody emptiness or absence, he had tried several times to calm Nitin, chanting formulas, but he had been caught the very moment he tried to take flight. Everyone was looking at him, eagerly awaiting his response. And Nitin wanted to do battle, aiming at catharsis. Up until then, Karan had played more or less well the role of the detached god, uninvolved in worldly affairs, unable to give more than a little comfort or confidence to his devotees. For him, God was nothing more than a kind of small machine to release some oceanic feelings. Any grievance was necessarily experienced as an aggression. Nitin forced this particular god out of his comfort zone, but curiously, what was exposed were the workings of the circuit that linked him to the deity.

When everything seems to be haunted by a profound crisis – power, action, responsibility, trust, legitimacy – when magic is no longer effective and because God, as he understood him, had withdrawn, no longer allowing himself to be so easily manipulated, all that remained for Nitin to do was to take up his own desire to act, in both its creative and destructive dimensions. He would find no one to make him *Chitragupta*, the great bookkeeper of the actions of men, or to enthrone him as *Yamadut*, messenger of Death.

After exhausting every possible path to an agreement, Karan and Nitin would reverse their positions. It was to be Nitin's turn to incarnate Bappa. But first, he did everything possible to teach Karan to be a 'good god'. He brought in a little disabled girl and invited her to ask Bappa when she would be able to walk again. Her mother asked Bappa to do everything within his power to make her walk again next year. Karan replied:

'Certainly. You must pray to Ganesh as much as possible and he will eventually grant your wish.'

But Nitin was not satisfied. 'Bappa, don't just say "certainly". It has to happen. There must be some result. There are many innocent children who are not able to walk. There are many children who do not have enough to eat.'

Karan replied that Ganapati Bappa always supported his followers. But Nitin was determined to tell Bappa about all the ills of the world today and to instil in him ways of responding to them.

'Rather than making an offering', suggested Nitin, 'in temples and mosques, it is better to feed a hungry person.'

The learning agreement was clear. Karan listened and took on board his informant's solutions.

'My aim', he said, 'is to inform you about all these things. That is why I have come down from Mount Kailash and am conversing with you all.' But Nitin was determined to not stop there. Whenever Karan took refuge in the posture of a blissful god, Nitin did everything to drive him into another state of discomfort.

'Although you are from Mount Kailash', he said, 'we have always lived here on Pali mountain (*Pali Pathar*). We don't care if you relay our problems to Mount Kailash! When we are residents of Mount Kailash [which also refers to the heavenly abode after death], we can't come back here, can we? As long as there is life in our bodies, we will fight to do something.'

And each time Karan clung to his staid old formulas ('Anything can happen if you want it to', 'Think hard and your wishes will be fulfilled'), Nitin introduced a new set of themes, just as disturbing to the god as the previous one.

'Bappa, I think my questions are confusing you,' he said.

'No, I'm thinking about something ...' Karan replied.

'Okay, you can keep thinking,' retorted Nitin. 'There are people gathered here, we would like to use this medium, the form you have taken right now, to spread good messages, spread useful messages to people through this good form of yours.'

'Certainly, that is my goal,' says Karan. 'That is the reason I came to this Earth.'

A man then approached to ask a question: 'Which errors should be tolerated and which should not be tolerated? Which misdeeds committed by humans should be pardoned and which are liable for punishment?' To which Karan replied, 'If you do a wrong, then you must also do a right. Sometimes we run over a snake by mistake while driving. In that case, all you have to do is offer a copper snake image to the god Shiva.'

It was then that Nitin, who was ready to give up and leave his spot for someone else, erupted with anger: 'This is blind faith! If I kill people by mistake, I am not going to offer an image of a copper man to God as a sign of repentance! If I kill someone by mistake on the road, I will not go to a temple to make an offering to erase my bad deed. I have to go to jail. Sir, I am a bus driver in public transport. An accident is the fruit of fate, but if I am involved, or any other person of our profession, I would have committed a punishable act. I cannot go to a temple and make an offering. I will certainly be put behind bars!' To which Karan replied: 'If you deliberately aimed your vehicle at the person, then you will be sent to jail, but if not, you will be innocent.'

Nitin Pawar had tried everything to control God's message, to make Karan say whatever he could. But without success. Karan then addressed the children, ordered them to repeat *Oṃ namaḥ śivāya* and asked them who was practising meditation among them.

'What is meditation?' asked one of the children.

'Meditation is something you can do at any time', said Karan, 'whether you are lying down, on the floor or sitting. You just close your eyes, palms up. And you take control of your mind. Whenever you do meditation, be careful not to let all your thoughts overwhelm you. You can increase your power of concentration by practising meditation.'

At this, Nitin flew off the handle: 'Bappa, kids these days are immersed in their cell phones!' he exclaimed. 'They go to bed and all they think about is texting. That's their meditation! Children today don't know how to chant mantras. There is not a single child today who knows a complete mantra. But tell a child to send a text message, he will be able to do it immediately. Kids see the image of Ganesha on their mobile phones and just say "Ganapati Bappa", but no one has any real devotion to him, and no one has the will to chant mantras. Facebook is their only mantra! We have a proverb in Marathi that says that man spends his life in the pursuit of pleasure and remembers God only when he dies.'

Karan replied that one was not supposed to start worshipping God only when death was approaching and that it was good to do it as young as possible. As for the use of mobile phones, he couldn't think of anything better to say than his usual refrain: 'Do you have the solution to this problem? It's up to you.'

In the face of a god who refused any commitment to the polity as he understood it, Nitin had first sought his approval to seek justice by himself. Faced with a man seeking empowerment, Karan had done what one would expect of a robot more than a god. Of course he was human, but he had often looked like a real robot, with whom a dialogue is possible, but a dialogue of the deaf because he jumps on a single word, never understands the whole sentence or else twists the meaning. At a twist in an argument, he would suddenly give the impression of understanding a lot more, or else differently, and the battle would start again, as though his incomprehension was only feigned, to lead the interaction elsewhere, to push it ever further.

Faced with this apparently unattainable god, yet whose coherence was only gradually being built, strengthening little by little his capacity to respond, Nitin showed patience, even obstinacy. Karan repeated what was said to him, establishing a semblance of agreement, before they slipped again into a new disagreement. Nitin's efforts to win him over were never completely in vain, but the Ganesh he had in front of him seemed to be inhabited by a principled resistance to being instrumentalized. Nitin had done everything in his power to make him the god he dreamed of, a god close to the problems of the people, an interface to propagate the good word, a pragmatic word, rich in practical solutions to solve questions of the common good. A worldly and modern god composed of an accumulation of good thoughts and good deeds.

The more Karan tamed his position, the less dialogue was reduced to a simple call to devotion. Yet, as soon as problems of the polity were discussed, the answer was always the same: he thought that people had to face up to their responsibility, that they had to stop accusing God, stop depending on him for everything and anything, hence his systematic side-step in order to take responsibility away from God, his denunciation of the 'magic' of rituals while promoting a form of mentalistic magic ('you just have to think really hard to get what you want'), and sought to make his interlocutors responsible. Nitin himself wanted to get out of the vicious circle, such that God would no longer be the one who is asked for anything and everything, and who is asked for more than he can give. He recognized that one does not accumulate 'karmic' merit simply by placing a coconut on an altar, but he could not envisage another possible pact.

How to get beyond the relationship of gift and counter-gift, the eternal balancing act between request and (dis)satisfaction? Unless, that is, the entity in question were merely a crystallization, an image materializing people's good thoughts, giving an 'image' of the way forward? But in that case, why bother with all the ritual machinery? If God is only an ornament, a decorative accessory, and not a crucial node of our actions, at least responsible for the results of our acts, then we should abandon him or rethink everything about our relationship to this entity *in retreat*, *invisible* and often *numb* that Karan personified so well.

'Bappa, if we have to do everything ourselves, why then should we keep you?' Nitin added. 'It's up to you to do whatever it takes to get things done,' the god said.

'I understand,' said Nitin. 'The actions to be taken are the responsibility of men. But aren't the consequences God's responsibility? Who makes the consequences of actions visible?'

Karan replied that God made them visible only through the actions of humans, and that the Indian constitution provided for different types of punishment for almost all misdeeds. Afflicted, Nitin tried once again to object: 'Can you tell how many people today follow the principles of the Constitution? How many people actually follow it? It is only followed in the courts!'

And as Karan insisted that humans were solely responsible for this situation, Nitin accused him one last time of shirking his responsibility: 'You keep saying that we are responsible for all the harm. If we are responsible for everything, then we are the biggest criminals! Therefore, we do not deserve the presence of the god Ganesha in our homes. We are not worthy of worshipping him.'

Karan ended the exchange with these words: 'If you are a criminal, you will be punished. You can lessen your evil deeds by worshipping Ganesha. [To the children] Children please ask me if you have any questions. Come closer to me. You can touch my trunk. I am listening to you. Come closer. *Oṃ namaḥ śivāya, Oṃ namaḥ śivāya* ...'

Nitin had been tricked by an autistic god. He had to repent for being 'the greatest criminal' and all he had to do was worship Ganesh to hope reduce the consequences of his bad actions! With Karan, the divine word appeared in its most naked form, like a meme or a 'psychic module' of self-control, combining self-regulation and self-generation of self-confidence (through meditation): 'If you think that [such and such] ... then all you have to do is worship me ...' Without thinking about it, almost inadvertently, Karan had run through a large number of possible postures. Sometimes detached or taking the position of an irresponsible God, at times he had also played the cultural idiot. Often empathetic, he had also defended the State against the weak, and this State god, hesitating between impartiality and indifference, had ended up becoming an accusing and redundant god in his vampiristic demand for unbounded devotion.

This latest bout of narcissism had awakened an impulse in Nitin that no one could have suspected. In turn, he had to become Ganesh. Karan had done everything to bring the situation to this point, he had stimulated in his interlocutor the desire to make his own god or to become one. Apparently, the misunderstanding was the great winner of the dialogue. But by dint of pushing each other and because they understood that there was something here that needed to be rethought, even sublimated beyond their incommensurable positions, Karan and Nitin had both given Ganesh a good reason to multiply, to yet again take another form. Nitin went to meet Karan in the small cabin which we had set up for him. He took the helmet and addressed the children:

'I am your Bappa. I am still alive amongst you. My name is Nitin Pawar. I am the Bappa of your region. My purpose is to touch your hearts and minds so that I can eliminate all your negative tendencies. Envy is a dangerous thing. Jealousy can ruin lives. If you want to overtake me on the road, despite the risk of an accident, that is greed. Traffic and congestion are caused by greed. If someone earns more than I do, then I will do anything to get rich or commit theft. Corruption is caused by greed. Taming greed is the ultimate solution. If I am poor, I must be content with what I have. If I have lower IQ, then I have to study accordingly. If I have a 125 horsepower motorcycle, then I don't need a more powerful motorcycle. Don't stop dreaming though. Your fulfilment should not come through harming others. Greed has undermined the worship of God. It perverts worship and devotion. It is the root of all our ills. Greed is our new God.'

One might have expected Nitin to say something else, something more revolutionary. Instead, he invited resignation, measure, self-control. The same interchangeable values that Karan had made his own. But it was not so much Nitin's words that mattered as their tone. Full of assurance and confidence, he had taken the place of the incarnant deemed deficient, as if to say to him: 'I will teach you to be a god.'

Note

1. 'What distinguishes technique from non-technique is a certain degree of circuitousness in the achievement of any given objective.' Alfred Gell, 'Technology and magic', *Anthropology Today*, no. 4(2) (1988): 6.

5

The Test

When Anand suggested becoming the incarnant, he had no idea of the suffering he would endure. He called himself an atheist, and, as is often the case in India, atheism does not mean that one does not perform all the necessary rituals. He had been present at the final moments of Karan's incarnation and had offered to take up and take on the work. He, too, had decided first to meditate, for quite some time, at the controls of the machine, learning from Karan's experience, letting out deep breath sounds, which inevitably attracted people's curiosity. When he opened his eyes, a group of young boys determined to test his divinity by any means necessary were facing him. He called out to one of them.

'Do you know who I am?' asked Anand.

'No,' said the child, who was called Rishabh.

'Am I different?' Anand said. 'Can't you guess my name?'

'I don't know ... [*hesitates for a moment*] Ganesh?' Rishabh replied, intimidated.

'Yes. I am Ganesh ... Ganesh. Then talk to me,' said Anand.

Another child came to Rishabh's rescue.

'Ask him if you are going to pass your exams.'

Anand pretended not to understand and made him repeat his question:

'Since you are God ...' Rishabh said, 'then give me the answers.'

Anand began to stammer before sinking into a deep silence.

Anand had been following much of Karan's experience and told us afterwards that at that moment he saw the same obstacles gradually building up before his eyes, like a video game. He was consulted like a divination interface, an oracle. But for him, miracles did not exist. And in his replies, he could only come up with banalities. Children asked him what their grades would be at school, others consulted him to find out if they would get married, or else which cricket team to bet on. After a few hours, Anand seemed overwhelmed. 'Am I just a prediction interface?' he asked himself, disappointed.

Anand's problem was very different from Karan's. Anand followed the family rituals imposed on him by his wife, but he was not the type of guy to meditate, and he could never have taken on the role of a ritualistic god, a role in which Karan had excelled, calling for the chant of *Oṃ namaḥ śivāya* at the slightest opportunity. In fact, Anand was lost. He did not know which god he might be or which god he felt like being. Gradually though, the experience of the experiment revealed it. The experiment revealed a curious god, both close to people and promoter of a wisdom not consciously known, always hesitating between atheism, anarchism and pantheism. This was the opposite of Karan, who had clung to ready-made formulas, proclaiming his divinity before launching himself into the arena. Anand had approached his incarnation with the opposite method. The question-and-answer game would dictate which god he would be. But at first, he wanted only one thing, to exit the machine. He felt trapped there, like a human who had arrived there by mistake in an envelope too big for him. Each question was a trap, a way to corner him. And he never had the last word. A woman approached him. She wasn't the only one to have the interaction get out of hand at first contact.

'What is your name?' asked Anand.

'If you are God, you should be able to guess what my name is!' she said.

'Do you know what my name is?' he said.

'No. What is your name?' she said.

'How come you don't know my name?' said Anand.

'Ganapati?' asked someone else.

'You should know my name,' said Anand.

'No, we don't know ... we don't know!' they all exclaimed in unison.

Anand assumed that everyone who came to visit him knew who they were dealing with. This was not the case. And, above all, many of them wondered how the device worked, suspecting that it had been rigged, or even that it was a perverse game to make people believe in the existence of a new avatar. The shrewdest could see that the robot was remotely operated, but could not determine where the operator was.

'I can be seen. I am in front of you. But I can't see you,' said Anand.

'You are hidden. We can't see you,' replied one of the children.

'There's someone inside,' said the other one.

'Are you hidden inside the apartment building?' asked another.

'I am inside as well as outside. Ganapati is everywhere. Some Ganeshas are seen and some are not,' said the god.

'Then move your trunk,' a little girl asked him.

'There is a hiding place for Ganapati in you too,' continued the god in a mysterious tone.

'Move your trunk, so that we know if you are Ganapati,' insisted one of the children.

'My trunk is already in motion,' said the god.

'Can you see the Earth moving?' another child asked.

'You are a ghost,' said another.

For some, Anand had to prove his divinity. For others, the cloaking of the operator made it possible to say rude things to God, or even to insult him or to play games that would finally push him from his pedestal. For, surprisingly, everything happens as though part of the users of the device took advantage of the fact that here god was caught within an artifice in order to take revenge directly on the non-existence of god, as if one reproached God himself for being an imposture. Two boys thus approached him with the plan to corner him.

'What will happen tomorrow?' one boy asked him.

'Past, present, future. I am everything,' answered the god.

'What will happen to me?' he continued.

'Only good things,' said the god.

'What's going to happen to him now?' said the boy, this time pointing to his friend who was with him.

'What's going to happen to me now, tell us!' repeated the latter.

'Nothing. Everything will be fine,' answered the god.

The boy meanwhile had slapped his companion.

'He has just been slapped, something bad has happened to him,' he said.

'Ganapati always says that everything that is going to happen is positive, it's not true!' said the other.

Because of his subjection to continued assaults, Anand had to become more inventive in the art of playing with paradoxes. A man approached him, showing that he took his aetheric appearance seriously, but with the firm intention of tricking him.

'Do you eat the same food as men?' he asked. The god said no.

'Then what do you eat?' he insisted.

'I don't need to eat,' said the god.

'How are you alive then?' continued the other.

'I am not alive,' said the god.

'Are you dead?' said the other.

'I am neither dead nor alive. I am immortal,' exclaimed the god, eliciting mocking laughter. 'I am alive in you.'

'Yes of course, we know that!' was then heard from the crowd.

'I am alive in you,' the god whispered.

Anand sometimes found formulas to get out of a tight spot, but his interlocutors were determined to test him on his knowledge of mythology and Anand refused to venture into that ground, looking for escape routes. Two students thus approached the robot with the aim of engaging him in a fight with no mercy.

'What do you know about Ganapati Bappa?' asked one of them.

'I know everything,' replied Anand.

'What exactly do you know?' the other insisted.

'Everything,' said Anand.

'What is Ganapati's first name?'

'Everything,' he said.

'What does Everything mean?' exclaimed the two students.

'Ganesh,' said Anand.

'It's not Ganesh. No Bappa, that's not the right answer.'

They then tried to corner him with another question, 'Why was Ganesha named *Ekadantā* (He who has only one tooth)?'

Anand often repeated the questions to gain some time. 'When was I appointed *Ekadantā*? Look ... why are you testing me like this?' he asked in exasperation. But they insisted. 'We want to know if you are real or fake. When was Bappa named *Ekadantā*?'

'Why then put me to the test?' complained Anand.

'First of all, answer our question!' they threw at him.

'But why?' retorted Anand.

'Because you are Bappa,' said one of them.

'Are you testing me, even though I am Bappa?'

'Because you know everything,' said the other.

'No, I don't,' the incarnant said, still trying to dodge the question.

'You're lying,' one of them then said. The other then returned to the issue. 'When was Ganapati named *Ekadantā*?'

Anand did not understand why they would ask him such a question if they already knew the answer. 'No, we want to hear it from you. It is a pleasure to hear it from the mouth of God himself.'

And in the face of so much obstinacy, he finally blurts out, 'But I don't know! I didn't write my own history.'

'Then who did?' one of the students continued.

'You wrote my story. You read about me. I didn't do it. I'm programmed. I only know as much information as has been programmed into me,' Anand said.

'If we wrote your story, you must have been programmed with the *Ekadantā* input. And since you are Ganesh, you must know everything,' insisted another.

'Who says I am?' Anand asked.

'Don't you know your own story?' one of the young men queried, bewildered.

'Please tell me,' implored the god.

'God has to pass every test,' said one of them.

'Why? Okay, but what if I fail?' the god replied.

'God never fails a test.'

Anand got stuck at every turn, giving the image of an ignorant god. His interlocutors took him for a god, when it suited them, and Anand in turn refused their terms. At the very moment he thought he could get through the test, inventing a robotic destiny for himself, he nevertheless had to prove his mythological knowledge. Fortunately for Anand, some people offered him respite, not seeking to question him, simply begging for his blessing.

'Who are you?' asked Anand.

'I am your devotee,' the person said.

'What is your name?'

'Niranjan.'

'Niranjan is a nice name. So Niranjan, what do you want to tell me?'

'I just came to greet you. I am very happy that you have come in this form. That is why I am bowing to you. I wish to have your blessings.'

'Ganesha's blessing will always be with you. Keep the same level of devotion.'

'I thank you.'

This kind of interaction, which had all the makings of an encounter with a talking idol, was rare. Both participants had to implicitly agree on who was speaking and where the voice was coming from. As long as these two problems remained unresolved, there was a suspicion of trickery, of deception, and therefore the impossibility of establishing a relationship of trust. And mystery was not necessarily the right solution, as with this child who systematically questioned any answer that seemed supernatural.

'Are you real? Where are you sitting?' he questioned.

'I am sitting in you,' said the god.

'So you are not a god,' the boy concluded.

'I am sitting in you. I am not the god. You are the god,' Anand went on to say. 'You are facing yourself. You are facing a mirror.'

His interlocutor was not convinced: 'No. You are sitting somewhere in the building. You are sitting somewhere in the building.'

Anand denied it and persisted. 'I am inside you.'

'Impossible,' the boy said.

'I am inside you', said Anand, 'or maybe I am outside you. I am everywhere. Or I am nowhere. I am abstract.'

'You are not a god,' insisted the boy. 'And you're talking from somewhere in this building. I know that. You are sitting somewhere nearby and listening ...'

The god persisted in his floating posture: 'Yes, perhaps I am sitting somewhere or perhaps I am not.'

'You are.'

'Maybe I am sitting in you.'

'No.'

'Where am I?'

'Ok, *bye bye*.'

Anand had finally invented an origin for himself. It was better to assume that he was not God but a man trapped in a machine. He thought this new scenario would greatly simplify his existence, but it didn't. Simply declaring himself to be a man was enough for him to then be sent back to the status of god, falling into another trap that was even more difficult to avoid.

'I'm locked inside this robot,' Anand complained to the man who was looking at him curiously. 'Locked in?' the man asked.

'Get me out of here. I am locked inside this robot,' said Anand.

'Since you are God, you must be able to get out by yourself!' the man replied.

'Why should we free you?' someone else chipped in.

'Yes, I am God. Are you going keep me inside?' asked Anand.

'How can we keep you inside? Since you are God, you can be everywhere,' the man said.

'No, God can't do everything,' said Anand.

'What can he do?' asked the man.

'God can't do anything,' Anand said. 'Everything that can be done is in your hands. God has given you limbs to act.'

'Then why do we need you?'

'I am not needed at all,' Anand concluded. 'I'm really not needed. I'm stuck here.'

Waves of interaction followed suit, the one after the other. Tests followed by showers of insults, followed by new waves of cross-examination for the attainment of divine status. The image of a man locked up, which then turns into that of a prisoner god who ends up confessing his impotence and uselessness, was no doubt conducive to generating debate, yet there remained suspicion, power struggles and difficulty in giving credence to each other's speech. It was only gradually that Anand understood how to avoid the traps that were systematically set for him. It was necessary not to make any claims, not to declare anything in advance, not to put forward a story line, not to proclaim his magical origins, not to want to show his omniscience, not to resort too quickly to paradoxes. It was necessary to be as little divine as possible, equal to his interlocutor, at least at first. For any figure that was asserted would be systematically thwarted. It was up to the people to decide who they were dealing with.

By the end of an evening in the old Mumbai neighbourhood called Kethwadi, Anand was exhausted. There were too many people to deal with, too much mischievousness, manifesting a critical spirit that had nothing to do with blissful devotion. The people of Kethwadi would not be dictated to by their gods, they liked to build up trials and tests. And every test made Anand realize his own naivety, the poverty of his good intentions. He had been pushed around and even humiliated at times, without being able to turn first contact into richer interactions. Even a question as simple as 'What's your name?' could lead to an interminable discussion without leading to a more meaningful dialogue. Anand had not been able to pass any test convincingly, either because he had not been able to grasp what it meant to be a god to these people, or because it was not possible for a man to pass such a test, at least not for Anand, because 'a god must be able to meet any and all tests'. No man can pass all tests. Of course, it all depends on the test in question, whether it is a question of answering a mythological quiz ('What is Bappa's first name?') or whether it is an answer to a metaphysical question that the interlocutor finds satisfactory. We have seen how a test can take very different forms. In all cases, Anand seemed to have tested the limits of his abilities.

A man then approached him with a question that was not fundamentally different from those he had faced previously. 'Is my friend, Somesh Sharma, going to be elected?' he asked. This time Anand did not answer. Tired of not being able to have a real dialogue in the way that he understood it, he would let the other person take up the deep reason for his existence, making public the very emptiness he was feeling at that moment. 'You know more about me,' Anand said. 'I don't recognize myself. You recognize me. I just want to listen to you. That's why I came here, in order to talk with you. To find out who I am.' And so the game of deconstruction began in the form of a reverse interrogation. Assuming that he was nothing and nobody a priori, Anand would perhaps find his reason for being.

'Who are you?' the man asked.

'You tell me. Who am I?'

'I don't know,' replied the man.

'Who am I?' the voice repeated.

'Ganapati, our father,' the man replied.

'You just said what I look like. But tell me who I am,' the voice insisted.

'Ganapati Dada,' said the other.

'Who is Ganapati?' asked Anand, feigning ignorance.

'He is our god of choice [*iṣṭa-devatā*],' the man explained.

'What is a god of predilection?'

'The one we revere without questioning,' the man retorted.

'Why do you worship him?'

'By worshipping him and practising meditation, we get a lot of energy,' the man explained.

'Energy?'

'Force, power,' the man clarified.

'Oh ... you get power. So you worship me to get power?'

'Yes ... to get serenity ... and power to act,' he said.

'Why?'

'It builds our confidence,' the man clarified.

'But who creates confidence?'

'I feel confident if the Almighty ...' the man started to say, before being interrupted.

'Who generates that confidence? Is it God or you?'

'God ... God,' said the man.

'Isn't it you who generates this confidence in yourself?'

The man appeared a little unsettled, yet he tried to answer the question as best he could. 'No, but I think that ...'

'Where is God?' interrupted Anand.

'It's a belief. It's a ...'

'Does God exist?'

'It depends on you whether you believe or not.'

'Does God exist?' pursued the god.

'He is if you believe.'

'What are you telling me here?' said the god.

'If you believe, God exists. If you don't believe, then he doesn't exist.'

'Isn't that what I'm trying to tell you?' Anand said.

Anand seemed to have finally found his style. He embodied an analytical Ganesh who subjected his interlocutor to hyperbolic doubt, hoping that a new apprehension would emerge from such provocation.

'So man is the creator of everything and God does nothing?' Anand resumed in an ironic tone.

'There is a kind of confidence that is created in us if we keep God in mind, a feeling that we can do it,' the man said.

'So this is an idea you form in your mind?'

'True,' replied the man.

'Is it really so?'

'Yes, all human beings have this power within them,' the man said.

'Hmm, that means you agree that God does not exist,' replied Anand.

'No, I don't believe that,' the man responded.

'Then?'

'There is someone, there is a force,' he said.

'What force? That power is in you, isn't it?'

'That's debatable because we don't know. When we are desperate ...' the man remarked.

'Why do you make things so complicated?' the voice said.

'I don't make it complicated,' the man protested.

'Simplify things. Just as a stone exists because you see it, great people also exist because you have met them, so God also exists because you can see him, right?' the voice said.

'No.'

'Why?' Anand asked.

'God is an inner feeling. He does not exist through someone's will or vision, he is self-generated [*svayambhū*],' the man said.

'We are saying the same thing. It is self-generating in you,' Anand concludes.

'I agree, but we need a support,' the man retorted.

The recurring problem of knowing where to locate God reasserted itself, although in a new way. It had not really arisen with Karan, who never stopped repeating that he came from Mount Kailash to the point of convincing himself. Karan, as a good divine robot, played on hypnotic dynamics. He played his origin myth on loop. Faced with such a redundant and narrow-minded god, it could only be said to him, 'well, since you think you are God, then I will ask you such and such a question ...' and, in the worst case, as we saw with Nitin, the bus driver, 'since you think you are God, then what is your solution to the traffic problem? And what about corruption?' Anand had realized the ineffectiveness of rote formulas. He did, however, have a plan. It is far from certain that he had it in mind from the beginning. It would be more accurate to say that it matured from one interaction to the next. It was a matter of convincing his interlocutors that God was within them, and not elsewhere. No one had taken him seriously when he had proclaimed it, loud and clear, as a way to solve the enigma of his presence in the machine. Each time he had said 'I am inside you,' the person he had spoken to had told him that he couldn't fool anyone. He had to either admit and accept that he was somewhere hidden nearby, or else invent a posture other than that of a god who magically appeared, coming down from a sacred mountain. He was too human and the cables were too visible for his presence to be attributed to a supernatural intervention, and he was too external, too material for people to accept the idea that he could be inside them as anything other than a poetic metaphor.

In order for the proposition 'God is in you' to be acceptable to the interlocutor, it was necessary that the question-and-answer game change in its very dynamics, that it no longer be a setting in which a desire to trap clashed with a will to foil interferences but, rather, that a kind of propositional game be set up, alternating propositions and counter-propositions, in order to arrive at a new proposition, or not, as the case may be, one not already considered by any of the participants. On reflection, the plausible alternatives for divine localization are not that numerous. God can be either external or internal, or else can navigate between the two in various regimes of appearance and presence. This regime of oscillation was for once the subject of a calm, well-argued discussion, far from the kind of conflicts that had taken place previously.

'Is a support really necessary?' asked Anand.

'Someone gave me ...' The man was looking for the word.

'A body?' the voice suggested.

'Yes, a body,' he nodded.

'But why? Hasn't God already given you a body?'

'No, but it is a human form,' the man said.

'You have already been given a body. Why do you need another one?' the voice asked.

'Yes, of course. But the feeling that I can do it. When you feel God in you, then you can achieve anything,' said the man.

This was a new interpretation of Ganesha, of his constitutive anomaly. It took an elephant-headed body to give human bodies the feeling that they could be all-powerful.

'But why do you consider God separate from yourself?' asked Anand.

'No, I don't think like that,' the man replied.

'Then you consider yourself a god?'

'No, if man thought of himself in those terms, then evil would not exist – but it does,' the man said.

'Why?'

'Those who have good intentions cannot commit evil deeds,' the man continued.

'But why do you associate God only with good intentions?'

'Because only He can do that. The Almighty only does good things,' said the man.

'Doesn't God create everything, good intentions and bad?'

'That depends on people,' the man continued.

'Isn't that the definition of God?'

'It depends,' said the man. 'It is an ideal. When I feel I have God in me, I do good work. When people think that God doesn't exist, that he has no power and that they themselves are all-powerful then ... bad things happen.'

A young boy decided that he was the one who would regulate the flow of the interactions, and he had decided that this conversation had gone on too long. 'Hey old man', he said, 'finish your conversation now, it's been an hour, leave some room for others! Time limit!'

For our interlocutor, the problem was not so much ontological, that is to say, whether or not Anand was Ganesh, but whether a form of wisdom could express itself within this mediumistic device, could come out of this tangle of cables, with or against its operator, no matter what the medium. And he was willing to do the test to see how much wisdom could emerge from the dialogue with this Ganesha-like yet unidentifiable voice. Even if that wisdom had to come from his own mouth and not from the god's.

Better still: it is precisely because the god was engineered that the man could dispense some wise thoughts on the status and function of the god in its relations with human beings. Impossible to simply incarnate God via Bappa, the device necessarily embarked the incarnant and his interlocutor on the task of chipping away and remoulding, as if the machine could not be just a simple transmission interface, one more hierophanic theatre added to the long list of modes of presence of the existing gods in a polytheistic context. The machine got up to speed only by going beyond the initial and troubled stage of *manifesting* incarnants, to become a *question*, a reflexive operator questioning the multiple ways for the god to exist in one and all.

After his tiresome initiation, we started up again, in the days that followed, with Anand as the incarnant. Anand insisted on repeating the experiment. He said that when he was in the position of controlling the machine, he had the impression that it was not he who was speaking. With such an interface, he thought, one was never completely the owner of one's voice, as if a voice other than his own was being heard. A voice paradoxically more interior and yet simultaneously more exterior. Divine ventriloquism and, as with all ventriloquism, 'the opportunity for strange voices to express themselves', which are not reducible either to the ventriloquist or to the one he manipulates.[1]

Obviously, it was not only those who addressed the god who experienced a kind of distancing effect in its presence. The incarnant himself felt it in his body and in his own words, which surprised him, as if they did not belong to him. Getting a grip by way of letting go was perhaps one of the reasons why Anand was able to resist better than Karan, who had clung as best he could to a prototype of a cheap contemplative god. Bappa was not a preformatted identity to be espoused, but a device of self-revelation. And Anand was determined to let flourish that third voice to which it had been so hard to give form. If God were to burst in here, it would be unwittingly, not because he commanded it or through an exercise of will, but because he would induce it of himself. An unknown god effectively self-generating, who found in him a favourable environment.

It was thus to this curious new entity that people addressed themselves, along with this new version of Anand. A less predictable god, one that was more attentive and who gained in confidence the more people that spoke to him. He was now being consulted on everything from family issues to public health, the pill, marriage, ecology, pollution, the benefits of music, politics, the necessity of rituals, and even metaphysical questions. Once the ontological status of the god was no longer an issue and Bappa was seen as a medium of wisdom, no matter how ambiguous the entity one was dealing with, another dialogue was rendered possible.

'Can you see me? Hello?' the man asked.

'Yes,' said the voice.

'My first question is this: the Bhagavad Gita mentions two types of yoga: *jñyana yoga* [the path of knowledge] and *karma yoga* [the path of action]. Which one do you think is superior?' said the man with a sardonic smile.

The questioner worked as a journalist for a local Marathi newspaper. He had waited a long time to ask his question, since there was a long line in front of the robot. Anand seemed surprised by the question at first and took some time to think before giving his opinion. When he was uncomfortable, he would stammer as if he was trying to make a point, but his stammering was never interpreted badly. It went unnoticed, as if people now tolerated that the voice of God is not given all at once, that it fumbles, searches and sometimes even gets lost.

Anand had succeeded in making people accept that he did not have all the answers immediately. He had managed to radically change the spirit of the dialogue by involving the interlocutor differently. The question-and-answer game was no longer a test of divinity, but an opportunity to share problems and to search for solutions together with a god who was never far away, floating between the machine, the operator and the interlocutor. In this configuration, the incarnant acted as a kind of indeterminate, fluctuating mediator who did not hesitate to say that what he said committed only himself, or to share his own 'beliefs'. The conversation continued on the question of the right track:

'Between the yoga of knowledge [*jñāna yoga*] and the yoga of action [*karma yoga*], which is superior? I think *karma yoga* is superior,' Anand said after much thought.

'So, my sub-question is, someone is a carpenter, another plumber, another doctor, do you think performing the rites is more important or is doing one's work superior?' the man continued.

'I think doing his job is superior.'

'Can you please justify?' the reporter asked the god.

'I can justify indeed. We are all limited in terms of ... The world is full of ... how shall I put it ... Some people may have the ability or faculty, or simply the opportunity to know more than others, which does not make them superior. The work that comes naturally to people, that's what counts. Ants and humans work equally. This does not mean that we are superior to an ant. We are all living beings. Each person who works and does what is right for him or her is doing a superior job,' the god said after a long hesitation.

'Okay. My second question is this: do you believe in Reincarnation?' the reporter asked.

'Yes, of course I believe in it, the child asked me the same question, sir,' replied the god.

'And in what we call eternal salvation [*mokṣa*]?'

'Yes, it is quite possible and achievable in one lifetime,' the voice affirmed.

'And which yoga do you prefer to attain this liberation, *jñāna yoga* or *karma yoga*?'

'All paths are the same, sir. You practise *jñāna yoga* or *karma yoga*, it will lead you to the same thing, salvation,' the voice said.

'Is it really possible to practise *jñāna yoga* in today's world?'

'Of course it is possible. Everything is possible, I think,' says the god.

'But *karma yoga* means doing your work with dedication and sincerity while *jñāna yoga* is something different, you have to be indifferent to everything that is happening around you and you just have to focus on the fact that God is eternal ...' the man clarified.

'The principle is the same. You have to attach and detach at the same time. You can't get too attached to something you can't detach from. That's the balance,' the voice said.

'Okay,' said the satisfied man.

Drawn into answering questions he would never have thought of, on metaphysical subjects of which he had only a vague idea, Anand had finally found the argumentative resources to get through it. His indifference to the alternative paths to liberation was not the same as Karan's disdain for worldly matters. It was, rather, a 'word of wisdom' that all paths are equal as long as one finds one's own. His interlocutor seemed sufficiently convinced to continue the dialogue.

'My third question', he asked, 'is this: do you believe in Non-Duality [*advaita*] or Duality [*vaita*]?'

'Again, it is the same to me. To me, everything is identical,' the god replied.

'Okay. So that means Non-Duality [*advaita*],' our man went on to say.

'No, I mean, it's also the opposite,' the god stammered.

'So you believe that God is present everywhere and in everyone?'

'Yes,' the god affirmed.

'Okay. So, what is your opinion, since you are Lord Ganesha, about the celebrations in Mumbai and Poona today?' the man asked.

'I went to Kethwadi and realized that people were selling tickets to see the deity. This is unacceptable, believe me. I have been to small villages and everywhere devotion is very strong. You don't have to bring money into this. It's a shame to have to pay to visit a god,' the voice said.

'So you don't condone preferential treatment to get around queues at the temple?' the reporter asked.

'Of course, I think it's a big mistake. I don't understand what happens in temples like Siddhivinayak. You can find Siddhivinayak in yourself, believe me. You don't have to go to a temple for that. I wouldn't have come in the form of a robot otherwise,' the god said.

'That's true. May I ask you one last question?' the man asked, pleased with the answer. 'It is a delicate question that concerns material things. I have been working for an association [*mandal*] organizing the *Ganesh Utsaav* for the last ten years. It is more a wish than a question. Whenever we go to companies to ask for funds, they say they don't want to interfere with religious activities. You see, I don't know what their policy is, but can you grant me a wish that next year the companies will sponsor us?'

'You want me to give you a guarantee that corporations will support you?' the god asked, seemingly shocked. The man laughed, a little embarrassed.

'Do you really need that?' the god continued.

'Yes, because right now we don't have enough donations to function. That has to change, that's my point of view, of course. You can justify yours,' the journalist said.

'In spite of all this, you manage to celebrate the Ganesh festival with minimal resources because of the goodwill shown by people. You should be proud as a community, trust your karma and continue to do so,' said the god.

'That was my last question. Thank you for answering this range of questions,' the reporter replied politely.

'Yes. You are looking in the mirror, sir. My words are only your voice bouncing back to you. All this is the product of your good thoughts. There is no god here,' said Anand.

'Thank you.'

'It is I who thank you.'

Recall the difficulties Anand had experienced earlier. It had been impossible for him to obtain the title of 'Lord Ganesha' without a modicum of ridicule or some caution that challenged him in his identity. No matter the amused tone of the interlocutor, he still showed respect. For the first time, someone had given him the title of Lord Ganesh after a series of questions, showing his satisfaction with the answers given. Then, crusading against the rituals and commerce surrounding the temples, Anand managed to find a new coherence by relocating the divinity inside everyone, including inside a robot. 'I wouldn't have come in the form of a robot otherwise,' he said.

The final part of the dialogue was surprising, for it showed that even an educated devotee, a journalist, a metaphysical enthusiast and the head of a neighbourhood association, could also want to obtain from God, in whatever form, even ambiguously, the granting of a wish. Anand was no longer there to grant wishes, his mission was elsewhere, but for the first time he could formulate it positively. That God, that cumbersome entity, should temporarily crumble, that the will to incarnate should fall to pieces, that there should no longer be any question of personifying anything, in order to overcome all relations of force, such that a true dialogue could be instigated. This is undoubtedly the profound meaning of his last sentence: 'There is no god here.'

This is something that may well seem trivial, but it is important to be trivial when one claims to be re-forging something. It is possible that Anand, in finding his own mode, has pointed to a much more generic, very simple mode of generating the divine through simple interaction. 'You are in front of a mirror,' Anand tells us. It is an interaction that follows the principle of any process of emergence (1+1 = 3), since it was necessary to let oneself be guided by the word of the other in order to surprise oneself and to express something fundamentally new, something which is better conceived on the model of the reflection of light.

Deep in the mirror, by dint of 'ping-pong' exchanges and answers turned into questions, a flash of wisdom must shine through, just as new for oneself as for others. This was a mode that Karan could not allow himself to experience, from the moment he considered that God already existed too much. It was necessary, then, to endeavour to imitate or to relay him. Thanks to Anand, can we now give a little more consistency to what sometimes shudders and sometimes rumbles under the concept of God? A magma of enigmas and unresolved questions. Bappa was the best way to track them down at the scale of the polity, sketching the contours of the polity by gathering all the enigmas, personal and collective, into a single apparatus of speech. Another way of doing metaphysics and politics at the same time, through a mapping of deeply held concerns. For Anand, Bappa was to be the means to make this journey into the abyss, to collect the mysteries that each person encounters in the course of his or her life, and to produce, via the machine, a collective awakening, by pooling all the individual illuminations of which it would become the receptacle. When he would eventually give his place to another incarnant, it would be to make other forms of wisdom spring up via the device, so that unknown sages who had never been given the floor could perhaps finally express themselves.

Note

1. David Goldblatt, *Art and Ventriloquism* (London: Routledge, 2006), 42. For a very nice theory of ventriloquism applied to social interactions, see François Cooren, *Manières de faire parler. Interaction et ventriloquie* (Paris: Le Bord de l'eau, 2013). The author develops the idea of interaction as the site of a continual 'dislocation of speech': 'All communication finds itself, to a certain extent, out of its time and out of its place ... because the ones interacting [les interactants] find themselves, in a haunted and ghostly way, staging figures

from elsewhere, who come to incarnate/represent/incorporate themselves in their interactions.' *Ibid.*, 59. Such dislocation is pushed to its apex here in our experiment. For a ventriloquial analysis of tele-operated robotics, see E. Grimaud, 'Pour quelques secondes de confusion ontologique. Échecs et réussites du contact dans une expérience de télé-robotique', in Thierry Dufrene and Joël Huthwohl (eds), *La Marionnette* (Montpellier: L'Entretemps, 2014).

6

Machine of Misfortune

An astrologer named Vishnu Modi had heard about the robot. He had applied for the casting, but was not selected initially, as we preferred to try first with Karan. Modi told us that he had the answer to all the problems we had encountered so far. He showed us a video of himself on the internet. He had an astrology show, answering people's questions live on TV after reading their horoscopes. 'And all this without any preparation, just improvisation,' he said. God had to be able to answer any and all questions: 'the whole package', as he said – and from all strata of society. Only an incarnate with this total ability to respond could claim to be Ganesha. It was clear to him that this exceptional human existed but that he had not been found. Even if he was right about this, Modi was irritating. He had a nasal voice and annoyed many of the production team. His astrology sounded cheap and he came across more as a shopkeeper than a *paṇḍit*. Nevertheless, he was full of enthusiasm, and at a time when we were unsure of how to proceed, we decided to give him a chance.

The apartment where we were staying was transformed into an astrological consulting room and the door was marked: *Ganesh Yourself*. People would come to consult a god who had a minimum of astrological knowledge, even if he wasn't a famous *paṇḍit*. Modi was so comfortable with the interface that he ended up doing consultation after consultation. It revolutionized astrological consultation, he told us. But against all expectations, Modi was not a politically correct astrologer. He had decided to ruin the lives of the people who came to him for consultation, giving them all sorts of problems. Even those who came in good faith and expressed their joy at finally being able to talk to Ganapati were met with the announcement of terrible misfortunes. We let him carry on. He seemed to know what he was doing. Never had a Ganesh been so disturbing to his devotees.

One day, a young artist named Sushant, the son of a poster painter, came to consult Modi. The two of them were placed back to back. This was sufficient to create a sort of invisible wall, and the artist addressed Modi naturally without being tempted to look at him. We realized that compared to a face-to-face interaction with an astrologer in a normal consultation, the presence of the robot introduced a kind of distancing or defamiliarization effect. When looking at the robot, the patient had the impression of talking to a voice coming from elsewhere. By putting the two people back to back, this effect was fully taken up. Sushant was aware of the nature of our project and was told in detail about the design of the robot and the reasons for the experiment. This did not prevent him from recognizing in Modi the incarnation of the god for whom he had been waiting so long.

'This is the first time that I have seen Ganapati and it is a special moment for me. I am talking to Ganapati Bappa ... I am very happy to see you. You are Ganapati Bappa!' Sushant laughs.

'It is because of your blessings that I am here, they have so much power,' Modi replied, before asking him his name, date of birth and the reason for his visit.

'I want to ask you a question on behalf of your followers. How do you feel when your creation tries to create and understand you?' Sushant said.

'How do I feel?' said Modi.

'You are in all your creations, aren't you?'

Modi nodded.

'And now your creation is trying to create you everywhere in every possible way, with plaster or electronics. And it tries to understand you and contact you in every way. How do you feel?'

Sushant wanted to know what God thought about the whole technological display that Bappa represented. And, it should be noted, he kept things well separated: there was God on one side and the interface on the other. There were questions he had been asking himself for a long time that only God could answer. And God answered.

'It's a great feeling, my friend. I have given many things to the world, I am in the water, I am in the trees, I am in the fruit, I am in your food, I am in your love, I am your father and mother. You try to understand me, who am I? What are my powers? I am very touched,' said Modi.

'I thank you for your answer. Another thing I would like to know, since we are all your creation, why are some creations inferior to others? Why should humans be superior to animals and plants?' asked Sushant.

'You are right. That difference will always be there. I can't balance these things. They come into being according to their *karma*. Someone is a pigeon, someone is a lion, someone is an elephant, someone is a human being. That's karma, my friend,' the god replied.

Sushant seemed satisfied with this answer. But he did want to know nevertheless how God could justify so many different religions:

'Very good question,' said Modi. 'All the sages of this world, whether *saddhus* or prophets, repeat the same message without you ever taking them seriously. We are all one, religions are illusions my friend.'

'Yes, we have created differences, that is the problem, our ego,' Sushant went on.

'You need to take this message to your circle of friends, to your family. Do you get rain on Sunday because you are Hindu? Monday when you're Muslim? On Tuesday when you are of another religion? I didn't invent religions!' exclaimed the god.

'Yes, we created them as human beings,' deduced Sushant.

'You are educated, you go to school, you learn so much. So why are you still not learning these fundamentals of life? Spread this message and you will flourish, my friend,' Modi continued.

Sushant showered him with praise. One could not imagine a better client for Modi. 'And your personal question? About your work, your finances, your health?' he asked.

'I don't know what kind of person I am,' Sushant stammered. 'I'm a good person, I think, maybe some things have come to corrupt me, but I'm somehow a good boy and rather wise. Will I be able to keep myself that way as long as I live?'

'My friend, that is very easy. I will tell you your problem. You are not able to express yourself at the moment. You have a very good heart, you have a very nice nature. But you have a hormonal problem,' Modi said.

'Really?' Sushant responded.

'You need to have a check-up,' said Modi. 'You have a medical problem related to your thyroid.' He added, 'Your body is your soul, my friend. You must take care of your body. Your health is very important.'

'Yes, health is important, but until now I have never had any physical problems,' Sushant said.

Modi insisted, 'I tell you, you have to check and on Friday you have to go to the Lakshmi temple and offer her a lotus flower. On Sunday, you must visit the Vishnu temple or the Krishna temple. I tell you, if it is not you, someone else will have a thyroid problem in your family. This problem will be solved if you do what I tell you, otherwise all your money will be absorbed by your health problems.'

'So, I have to take care of my health ... I thank you. Long live Ganapati,' said Sushant.

'Welcome, my friend,' Modi concludes.

How could the astrologer have allowed himself to make such a prediction? His diagnosis was astrological in name only, since he had simply asked Sushant for his date of birth, but had not studied his birth chart. And how could he not have realized the implications of saying such a thing? Modi looked like a bad psychic, a charlatan with bad intentions, abusing his position of power to announce disasters and enlist his clients in devotion. Every consultation with Modi ended with the suggestion of a simple remedy: go to the temple to avoid any form of calamity. Although such abuses seemed to bring the astrological profession into disrepute, Modi was experimenting with a new divine posture. In order for it to exist, it had to scare people. And he was really convinced that going to the temple was a solution to everything. When the consultation was over, Sushant explained how he had experienced it. He did not feel that he had been taken advantage of. Instead, he had experienced the presence of a distant, non-judgemental being, not Modi but God himself.

'It's really very different when you talk through this machine than when you're in a regular consultation,' Sushant said.

'You were close to crying, weren't you,' Modi pointed out.

'I really thought you were Ganesha,' Sushant continued, moved, 'because normally when you go to the temple, you can't talk. With the machine, you can confide your deepest feelings and ask the fundamental questions that are bothering you. I began to think, without knowing it, that this was indeed God. Humans can judge you, point out your weaknesses. It's hard to talk to any other human being. Our friends can be very critical of you, but with this, you know this isn't going to hurt you or judge you.'

'Great!' chimed in Modi.

Surprised that such a feeling was possible, after the hectic interactions we had had so far, I asked Sushant if he had not felt at any point that he was talking to Mr Modi.

'No,' he said. 'To me he was more like Ganapati Bappa. I am not face to face with him. He's somewhere behind, and what I was seeing was more my idea of Ganapati Bappa. So, for me, there was no doubt about it. I was asking questions to Ganapati Bappa and not to any particular person. What was important was Ganapati Bappa, it was he who occupied the space ... Because you can't talk to a statue when you are worshipping it, you give it flowers and make offerings.'

'At the same time, you knew you were talking to Mr Modi, didn't you?' I insisted.

'He was not in my mind. I thought I had Ganapati Bappa in front of me and I was concentrating on what I had to ask him. I wanted to ask him more things, but it's not easy,' confided Sushant.

'Super hit!' said Modi.

The astrologer was proud of his success. But the truth is, he had nothing to do with it. He had only served as a medium. Sushant was not a particularly naive guy. He had simply managed to abstract himself from the figure of Modi, to expel him from the interface, to speak to 'his idea of God' in a pure face-to-face encounter with a presence from elsewhere. There had been no ontological confusion here between Modi, the machine and God, but a curious phenomenon of initiation by which Sushant had felt the spectral presence of a god suddenly become so close. This explained why Sushant did not mention Modi's warnings about his health at all. The astrologer as a character had disappeared for a few minutes from the interface. Sushant had not felt any torpor. On the contrary, the consultation made him joyful. Given the ease with which he had managed to persuade himself without any effort that he was talking to God, he felt that the machine could help many people to express themselves. Modi confirmed all the ways he could benefit from it.

'For counselling, for problems of depression, it's very good,' Modi remarked.

'People will feel more comfortable with Ganapati Bappa than with a psychiatrist. Science and God together, what a miracle!

All this to take away someone's pain ...' exclaimed Sushant, full of enthusiasm.

'How long was the moment when you felt like you were talking to Ganapati Bappa?' I asked.

'Actually, when I was talking about Nature. Here was something I really wanted to ask God. Why are human beings so dominant over all the other things that God created? They monopolize everything. That's a question I really wanted to ask God if he was around. And so I did,' he said. And then he turned to Modi. 'Actually, you weren't important. I don't feel like it was you who was responding to me. It was someone else through you. And it was coming from far, far away. And because of the interface, the effect is multiplied.'

'There was a lady who came to thank us one day for designing this machine,' I said. 'She said, "This is the first time I have seen my wish fulfilled by a god, a robot and a human at the same time." This meant that she thought she was talking to three people: a god, a machine and a human. But for her, there was no contradiction and this was not a reason to disqualify the interface, quite the contrary. The god, the machine and the human were brought together in a virtuous combination. Did you have the same impression?'

'I took it as one thing,' replied Sushant. 'I didn't think I was talking to a person. I wanted to ask a question and get the answer. And maybe it was in me, I have wanted the answer for a long time, that's why I rushed to ask it. I didn't think about who was behind it. I had this question to address to Bappa and I asked it.'

'You were totally taken in the question, the concentration was there. You had tears in your eyes ...' Modi continued.

'The question was emotional. For me, it was a bit like a confession,' said Sushant.

'It will be a great success, no doubt,' said the astrologer.

'If you keep the machine in one room and the questioner in another, I think there will be a big line,' suggested Sushant. 'That's divine psychology! We won't think about the operator, but it will take some convincing.'

'Super hit!' Modi said again.

By suggesting that we put the machine and the questioner in two separate rooms like a Turing test, Sushant was unknowingly urging us to take a step backwards. We had realized early on how hiding the operator only created confusion and mistrust. For trust to be established, it was necessary, on the contrary, that the incarnant be clearly visible, that the interlocutor be able to reconstitute the circuit that leads from Ganesh to the operator. And paradoxically, the less the system hid its components, the more it was up front about its relational nature, its ways in and ways out, the more the questioner could abandon himself to the experience, in the mode of 'as if it were a God', at the moment when he put forward his question, to have the feeling, in the end, that the answer came to him effectively from elsewhere, and perhaps from God himself.

It had clicked in a dramatic way in Sushant's case. He had found the answer to his metaphysical question satisfactory. He was touched. The rest, starting with Modi's catastrophic predictions, was of less importance. It is known that in the arts of prediction, any word can hit its mark, no matter who utters it. But this one was meant to lead Sushant to the temple, which he already visited regularly. In his case, there was no reason to be affected by it.

Modi increased the number of consultations. But it was Sushant who had been the most committed. Those who came after him were more sceptical, and their resistance was much less easy to break. One of them, Shailesh, was sceptical from the start. He said he did not believe in astrology or God. And for him, Modi was an astrologer trying to pass himself off as a god. Nevertheless, Modi managed to draw him into one of the most surprising dialogues we ever witnessed.

'Greetings Ganeshji. I came to see you because my friends told me that you give advice to people about their lives, their future,' Shailesh began.

'Welcome my friend,' Modi said.

'Just to be very clear', said Shailesh, 'seriously, I don't believe in all of this, I believe in you, but not in anyone who can tell me what my future consists of. I believe that I am the master of my own future and I lead my own life.'

The arrogance displayed by Shailesh at the outset displeased the astrologer. He asked him why he had come to see him. Shailesh told him that he was really happy with his life, that friends had told him about this consultation device and that he was curious to try it out, to get some tips on how to make his life better. Modi told him that he had not appeared among humans by magic spell, but for a reason, that it was the love of people that had summoned him here and that his mission was to help us all. He was trying to sound like a great enlightened guru, but it didn't ring true, especially since he had an unfortunate tendency to stutter.

'So what do you have to say about my life?' asked Shailesh.

'What are you doing right now? What is your name my friend?'

'Shailesh Rai,' he said.

'Shailesh my friend, listen to me, put an end to your ego,' said the voice.

'Okay.'

'It's not getting you anywhere. You are not impressive at all.'

Shailesh initially looked taken aback by such a diagnosis, before pulling himself together. Let us remember the dialogue between Nitin, the bus driver, and Karan, which had been deeply confrontational. Forever in a position of weakness, Karan could not tackle the enormity of the problems that were submitted to him. But he turned a weakness into a strength. For him, man was solely responsible for his misfortunes and the power of God would reside in ever greater withdrawal. A whole metaphysics is embodied in an attitude, a style, details of elocution, a way of speaking or of not wanting to speak. Fleeing, always preparing the conditions of his disengagement, Karan, in his very attitude, had marvellously embodied the withdrawal of the divine to the high spheres of the cosmos. Withdrawal from the conversation he wanted to escape, withdrawal from the problems of the polity that were a little too heavy for him to bear.

Like a living icon, he had been able to do only one thing, to encourage people to meditate, and all he really wanted to do was to dilute himself in the depths of the universe, to disappear into an infinite silence or to melt into a great oceanic vibration, the only truly comfortable position for God. Between Modi and Shailesh, it was yet another psychological war. The incarnant began by attacking Shailesh, who, caught by the throat, had no choice but to endeavour to defend himself. From this change of posture, another vision of the divine would perhaps emerge. It was necessary to exhaust the figure of the authoritarian god in order to examine what powers and displacement of qualities and properties were at work, just as Nitin and Karan had previously exhausted the figure of the benevolent and responsible god in order to find in their power relationship a connection with pure energy. Modi had not said anything to us about his strategy, yet I cannot explain the angry mood with which he started the conversation with his interlocutor in any other way:

'Stop being so recalcitrant!' the voice said to Shailesh.

'I'm not,' Shailesh retorted. 'The first thing in my job is to always think positive and act positive. I never say no to anything. So how can you expect me to be negative or pessimistic? I always say yes to everything in my professional life. That's what I'm supposed to do. If someone tells me they need anything, I make it happen within a maximum of six hours.'

Modi changed his tone and began to be downright contemptuous, 'This is your job my friend, there is no pride to be taken in it, you get paid for it, you have a salary, it's not something great that you accomplish there, have you no other purpose in life than that?'

Modi's technique wasn't so bad, as Shailesh began to reveal aspects of himself after just a few seconds of interaction. 'I don't want to do anything but work and make money. Everyone in this world runs after money, and me too.'

'What about your image, don't you care even a little about your reputation?' the voice asked.

'I guess I have a good reputation,' said Shailesh. 'People call me for more and more work. I don't chase people. I've earned a certain notoriety.'

Divination professionals know this very well: if someone comes to see you, making a show of their scepticism, you have to go after them, prick them where they might be vulnerable. And it didn't take long for Modi to make a decisive breakthrough in the psychology of his interlocutor:

'Okay, now you want answers, some good advice on how to succeed in life, right? First of all, I want you to take care of your mother and your family. Your family, especially a lady, is very upset right now. You are not taking responsibility for your family. Secondly, there is a lot of negativity in your home. You are not taking care of your finances. You are not a responsible person.'

Shailesh wouldn't let himself be pushed around, and responded immediately:

'I would like to answer this. No, I take care of my family. And yes, I don't have to give money to the house, because my father still runs the family and he doesn't ask us to contribute. We are three brothers and we don't give a single rupee to the family. And about managing my finances, again, you are wrong, because I bought a lot of properties with the money I earned. So, I don't know where you got that idea. I've invested a lot. How do you think I fail to manage my finances or fail to take care of my family?'

In a few minutes, Modi knew almost as much as if he had done a lengthy investigation of the person he was talking to. The man in front of him, however, was ready to defend himself.

'You won't know until later,' he replied. 'At the moment you are well, but the wind will turn badly for you my friend. You will have health problems related to your stomach and your bones. There will be accidents, fractures and you will have blood problems too.'

This cascade of unfortunate events did not overawe Shailesh at all.

'Medically any tall, thin person will end up with bone and joint problems, so I don't think there's anything new in what you're telling me that I don't already know,' he said.

'No, no,' Modi replied. 'You're absolutely wrong. I don't care how thin you are or how tall you are. I read your horoscope, your *karma*.'

The distance produced by the robot had a different effect on everyone. Those who perceived that there was an operator, at a certain distance, allowed themselves to be more aggressive and critical than in a face-to-face interaction with a human. For those who saw the Ganesha trunk before they saw the operator, the divinity of the thing acted as a brake on their aggressive impulses. One had to be respectful. Shailesh confided to us that he belonged more to the second category. Ganesh's trunk had bothered him, because he was not ready to accept just anyone like that, and he would have preferred a confrontation with an unmasked Modi. For his part, Modi made use of fear to cure his interlocutor of his arrogance. Family, money, illnesses, everything was a pretext to make his victim feel that one day he would be overwhelmed by uncontrollable events. Modi was determined to reduce the ego of his interlocutor until he understood that he was only a small particle in the universe. And when Modi said the word *karma*, Shailesh's face became tense with concern.

'Okay, but, karma, it keeps changing, right?' he asked.

'Yes, it changes all the time,' said the astrologer, sure of himself.

Shailesh could have told Modi, if he had been consistent with his initial posture, that horoscopes or karma were not things he gave much credence to, unless, at that very moment, he had a doubt. However, the exchange actually branched off in a much more interesting direction:

'Are we talking about fate or luck?' Shailesh asked. 'Being God, I don't think that you should talk about luck with humans.'

'Why not? Luck is a very important aspect my friend. You are lucky. That's the reason why everything is going well for you financially. There are people who are not lucky, who work like animals, who earn 100 rupees and don't even have a house. They live on the edge, they work harder than you. Luck is because of your karma, you did good things in your previous lives, that's why you can enjoy this life. You have a home.'

Shailesh could not but agree with this interpretation. Nevertheless, he sought to point out a contradiction in the word of the divine:

'I was lucky enough to be born into a well-off family, who gave me a good education, and that education allowed me to live my life the way I wanted. I had that fortune, yes ... I agree. I have my parents' house, my own house, which is why you contradict yourself when you say I don't know how to manage my finances.'

'You'll see later on my friend,' Modi warned him.

'But it's contradictory, you were talking about my finances today, I told you that everything was fine and now you're telling me that the future will prove you right! You went from the present to the future!' protested Shailesh.

'Absolutely,' the astrologer replied.

The exchange shifted a notch in aggressivity. Shailesh was not about to let such a bleak future be predicted for him:

'I'm telling you that all these things are not a problem for me in the present. I manage my money very well, my family is happy. My mum is a little worried about my older brother, but nothing concerning me.'

Modi insisted again and again:

'No, my friend, now listen to me. A big problem is coming. It will affect your finances, your health, legal worries, it will affect your brother too, your family life will be affected too.'

To defend himself, Shailesh could only point to the generalities of the diagnosis offered to him:

'Anyone can encounter these difficulties you list. Why should they all strike me?'

'No', replied Modi, 'it depends on each individual.'

'But you have mentioned all the different problems that can affect anyone in life,' Shailesh pointed out. 'There aren't any others: finances, family, brother and mother, legal problems. What's left? Nothing. The only thing left is death! So, with the exception of death, all the problems that exist in this world are going to strike me?'

Faced with this possibility, Shailesh's voice had changed tone, a touch ironic; it let slip a bit of fright all the same.

'You don't understand what I'm saying,' Modi told him. 'You will have to solve this problem. No one is going to help you. You must find a solution yourself. You will be alone my friend. Understand what I am saying! I am not talking about everyone, I am talking about you, I am focusing on you, you will be alone, hear what I am saying!'

'You mean my family is going to abandon me in that moment?' retorted Shailesh who looked really affected this time.

'Yes. You will have a lot of difficulties, you will have a lawsuit against you, you will go to jail, you will have problems with the bank, debts, it will be a vicious circle.'

One cannot blame Shailesh for giving in to this fear-mongering god. But Modi apparently did not expect his defence system to fail so quickly.

'Can you give me a specific time frame? Ten years, twenty years, thirty years?' asked Shailesh.

'Ummm ... Five to seven years from now,' Modi said, after some hesitation.

'You mean I'll have to take out a loan or something?'

'Yes, you're going to have an obstacle in your path.'

'Until today I have never taken out a loan. I don't believe in borrowing money,' said Shailesh. Modi then had a surprising response: 'I'm just giving you an idea of the kind of problems you'll face. My friend, I know you don't believe in astrology, but people come to me, they ask me, "God, what is my future?" "God, what should I do?" "God, what will become of me?" "God, what will the future have in store for my children?" You created astrology, not me!'

Modi felt held hostage by his role as a soothsayer and expressed this strongly. Once again, the device revealed its reflexive virtue, its *meta* character. The astrologer had not engaged his interlocutor in a divinatory consultation like any other, but in an emblematic interaction that had to contain within it all other consultations.

'Yes, I agree with you, my parents are believers for that reason,' Shailesh pointed out. 'They believe in all these things. I'm the only one in my family who doesn't agree with all that.'

'But you have to stop them too', said Modi, 'because they come to me and I have to fulfil their wishes. Do you know what the meaning of astrology is?'

'I don't think God created astrology,' said Shailesh. 'My parents go to God to worship him, not to know their future.'

'Astrology means *Time machine*,' declared Modi. 'It turns back your time and it is machine-like.'

There was certainly something atypical in this interpretation, but also about Modi's confession that he must be as insensitive as a machine. Once again, the machine had not been very present in the exchanges. The technology seemed to have been forgotten, in appearance at least. Because we will see when the two interlocutors share their impressions with us, that it was never completely forgotten. There was no doubt that for Shailesh, Modi had oscillated between two possibilities: being God and/or the astrologer that he was, playing on the confusion of roles. Ready to acknowledge him as a god, he nevertheless kept denouncing the fact that God is not there to make predictions. But Modi had oscillated between at least two other possibilities, empathy and coldness. The mechanism of telepresence allowed him to test very concretely the possibility of being an emotionless machine-god and to depart from the empathy required of an astrologer in an ordinary face-to-face consultation.

The exchange was not over, however. Shailesh wanted concrete solutions. 'Since you've outlined all the problems I'm going to have in the near future, ten years from now, is there a remedy?' he asked. 'Excellent question,' said Modi, a little surprised by this new twist. 'If I prescribe a remedy, will you follow it?' Shailesh replied that if he thought it was possible and healthy for him to do so, he would certainly try. At this point, Modi became so confused with his instructions that he bordered on the ridiculous. 'On Saturdays, my friend', Modi said, 'you must make a visit to the god Shaani and another to the god Hanuman, okay? And a visit to Ganesha every Monday. Sorry, a visit to Ganesha on Tuesday and another to the god Shiva on Monday ...' Shailesh pointed out to him that during these three days, he imagined that he should not drink, that he should be vegetarian and abstain from sex. Modi told him that sex was 'love' and had nothing to do with his horoscope. Shailesh insisted on remaining carnivorous on those days, because for him, eating meat too was 'love'. By going to the Shiva temple on Monday, the Ganesh temple on Tuesday and the Shaani Dev and Hanuman temple on Saturday, Shailesh would change his life. 'All these remedies will make you happy,' Modi told him. 'You will have all the solutions to your difficulties, you will acquire more and more properties, money will shower your life, your marriage will be wonderful, you will have great children!'

Shailesh had defended himself, he had tried to hold his position of the absolute sceptic for as long as possible, pointing out blunders and contradictions when they presented themselves, but the astrologer had finally made him crack, through predicting the worst horrors about his future. Shailesh even confessed his love of certain gods. 'You know, I have a lot of respect for Shaani Dev or a saint like Sai Baba,' he said. And the discussion shifted from belief in God and trust in astrology to the need to make the right choice, to make the right contracts, not with God in general, but with gods that are 'tailor-made' for us. Shailesh had to identify his gods of choice. This was the ultimate trick Modi had used to his advantage. Even for someone who claims to reject God, the gods in the plural are always there, it is enough to identify the good ones, and if some of them turn out to be ineffective with practice, there is nothing to prevent trying others.

Modi didn't have to do anything to trigger divine love when it came to Sushant. Sushant had happily accepted the potentiality of having Ganesha in front of him. Receptiveness was perhaps the best defence, for the threats of the astrologer had been perceived by Sushant as painless, lost in a stream of possibilities, in the mood of 'why not'. Shailesh, on the other hand, had been sensitive only to the frictions and inconsistencies of the incarnation process, pointing out every occasion when Modi, God and astrology were not quite in harmony. For him, Modi was an impostor. By dint of struggle, however, he had finally given up. He even asked when he would start to feel the changes if he followed the remedies prescribed by the astrologer. From what he had heard from other people who had consulted and followed the directions carefully, 'you feel inside that something has changed around you'. Modi gave him fifteen days. Shailesh thought that a test of three months would be better, because he didn't think that God could have any influence so quickly or that his situation could change in such a short time. 'Yes, after three months', said Modi, 'it will be wonderful, I tell you, you will not stop, you will continue and continue forever!' 'Thank you very much Ganeshji,' Shailesh said in the end.

Are we to conclude from this strange defeat that only the flow of speech matters, no matter what it carries, that predictions uttered by a cheap astrologer can affect us so much? No matter who Modi was in the end, Shailesh could not have predicted that one day his defences would crumble one by one as he faced the worst prospects. Although he would not, in the end, apply Modi's remedies, he had considered doing so, because the dialogue with the machine had stirred up an old, carefully buried core of anxiety. If God survives to such an extent, it is there at the limits of our powers of calculation and anticipation, at the blind spots of all predictions, coiled in the heart of uncertainty, in a slight temporal shift, where humans are simply unable to project themselves, to understand what happens to them, to explain what has happened to them or to guess what will happen to them.

But we must go further, because there are multiple ways to occupy this gaping hole. Because of his practice as an astrologer, Modi had perfectly understood a movement that was difficult to grasp, a very particular way of making the god emerge by taking from his interlocutor a piece of time, of his time, to make of it an *active power*. This time, which Shailesh cannot anticipate, because it is impossible for him to act by taking into account all probabilities, Modi throws it back in his face in a personified form, one that is easy to grasp. It's another way for the god to emerge, which contrasts with the previous ones, notably with that of Karan, who believed in the power of repetition but exhausted himself in a pale copy of the divinity after having proclaimed himself 'electronic Ganesh'. It is a very interactive way of generating the god. Anand probably introduced us best to it by realizing that it is in dialogue, by learning to identify where his interlocutor situates his own zones of uncertainty, that God can be relocated, in himself, elsewhere, redistributed everywhere. But Anand had a static apprehension of this problem, more spatial than temporal, whereas Modi, as a good professional diviner, understood that divinity can be generated from time.

This brings us closer to, while shifting away from, the modality of existence that the *paṇḍit* in the first chapter had allowed us to identify: any natural form of the god (stone, root etc.) that has not been generated by the hand of man is said to be 'self-generated' (*svayambhū*). What should we call this form generated from a line of time? A retroactive god? Where Karan had failed, thinking that it is by repeating the same hypnotic formulas that one becomes a machine god, Modi had only applied a basic principle of cybernetics: to create, using the uncertainty of the being he had in front of him, a feedback mechanism, a feedback loop. This is probably the reason why the machine suited him so well and why it was wiser not to entrust it to him for too long. For Shailesh, it was clear that we should not leave the robot in the hands of the astrologer. Hijacking the machine for divinatory purposes, he felt, was dangerous. But, in reality, Modi had simply succeeded in making himself for a moment the master of time, generating a shadow god, without Shailesh noticing, from the very flow of his very own words.

7

The Matrix

Modi was constantly harassing us. He told us he had gotten approval from his network for the robot to appear on television, on one condition: that he was the embodiment. He made himself available for consultations whenever there was a gap in the robot's schedule. Shailesh did everything to spot 'wise men' who could spread the good word. We tried with a friend of his, a feminist journalist, but nothing much came of it. Anand also went looking for potential incarnants. He too sensed the danger of letting Modi take over the machine. He knew how difficult it was to incarnate Bappa, and was quite willing to share. So the 'casting' never really stopped. Karan was also always there, ready and willing to do an incarnation when needed. Other people who had tried the experiment spread the word, and gradually Bappa had its own pool of operators to draw from as circumstances dictated.

This pool grew daily. Mumbai was our field. The robot was taken to the four corners of the city when someone offered to give it a go. We offered two types of experiences to the incarnants. Either an open-air interaction with an audience, knowing the risk that the question-and-answer game entailed. Or a more intimate interaction where the incarnant had to put himself in the place of God and answer our questions or choose to address an audience and deliver a message as if he were God himself. We didn't plan on broadcasting it, but everyone hoped that one day what they had to say would be heard, because putting oneself in

God's place, we realized very quickly, made people say things they would never have expressed otherwise. *Ganesh Yourself* meant expressing loudly and clearly what one had to say, setting one's message in motion without worrying about the recipient, away from any public, intimately, looking for the purest transmission, hoping that what one had to say would be relayed by the cosmic waves.

Prior to the transmission of any message, however, *Ganesh Yourself* meant an experience of incorporation, one that is difficult to comprehend until you experience it yourself. The mere fact of putting on the operator's helmet and seeing oneself equipped with an elephant's trunk first of all changes one's self-image. And if one then plays the game of being God, then it really is possible to be surprised by the metaphysical content of one's own words, as if they were emerging from the depths of the subconscious or were dictated by a voice of a higher order.

Jyot, who applied one day to be an incarnant, is a good example. Once Ganesha's helmet was donned, the young woman's large eyes and her voice gave the god a new appearance. No one had thought that Bappa could be a woman. But Jyot loved women and was convinced that Bappa must go beyond gender boundaries. She came to the casting and returned several times to try it out. And little by little her voice became smoother and smoother and ever more hypnotic. Prior to any attempt to conform to the norms and standards of divine speech, the encounter that follows helps to identify how the mainsprings of an organic transformation operate. While Jyot was trying to control her eye movements in order to look at the person in front of her, we had a dialogue that I cannot resist transcribing because of the richness of the lessons it contains.

'Am I looking at you?' asked Jyot.

'Ganesha is not like the Mona Lisa,' I said. 'Mona Lisa looks at everyone, Ganesha's idols look at no one in particular, the eye always slightly above.'

'I see', replied Jyot, 'but it might be looking inside itself too. Or be my projection, of what's going on in my head?'

I made no secret of my admiration for this new incarnation being born: 'You would make a wonderful god, Miss Ganesha! How do you feel about your trunk?'

Jyot seemingly caught up in the game, answers me in a smooth voice, 'I like it when it moves. It takes me by surprise too. I like the wires that stick out. But whose idea was it to add these frills ... The curious effect of this trunk is that you suddenly feel responsible for what you say!'

While Jyot was busy taking control of her trunk, I thought to myself that I had never seen an incarnant with such an animated face until now: 'You are the most expressive person we have met when you move your eyes ...'

'Yes, my family didn't like that at all, no one appreciated my talents during family dinners.'

'Can you do the *Navarasā* [nine emotions of classical theatre] with your eyes?' I asked the incarnant.

'Is "psychotic" one of the nine emotions? Can you tell me if there is a difference when I look far away or close up?' Jyot gives the impression of squinting. 'Yes, a convergence effect. Now, how do you feel Lady Ganesha?'

'I like to swing my trunk,' she says, laughing.

I point out that the sensors react randomly and she can't control it.

'That's what you think, I'm God, I control what I want. In fact, it's not about self-control but rather about letting go. It's about being out of control, you understand ...?'

'You mean maximum *self uncontrol*?' I said.

'In fact, God is not even imaginable to you human beings. So you should be happy with what you think you understand, but don't try to understand everything, because there is no point.'

There is laughter.

Jyot had a lot of spirit, much more than many incarnants. Above all, she confirmed that a metaphysics can be embodied in small details and not just in speech, in an attitude, a posture, a look, a letting go. All this contributed to make her a singular Ganesh, a medium of embodiment of a conductive texture for an idea of God different from the others. 'To use a phone, you don't necessarily have to understand how it works. It's the same with God and this universe. It's beyond your understanding. But it's there. You don't have to focus on Him for Him to be in your life. It's best not to exhaust yourself trying to understand all its facets, because it's impossible. In other words, the less you think about it, the more He is there. Living a life connected to God is better than striving to unravel his mysteries.'

Jyot had made the interface her own. She formed with it a curious crystalline assemblage, inhabited by a bright and provocative light comparable to that of the great mystics. 'Do you accept your limitations as a God? You can't move your ears, that's a great handicap for an elephant ...' I said, in a playful tone. 'It depends on your perception and what is meant by movement. You only see through your access to space and time. The movement in my ears is so vast that you can't even detect it!'

Jyot had found her style, a mysterious and paradoxical deity. 'Fixedness of mind is the ultimate movement. It is what all human beings ultimately strive for. Can God be evil? What do you think?' she asked me.

Jyot had a subtle and magnetic way of engaging her interlocutor in enigmas, of handling paradoxes, of leaving questions unanswered. We were far from Modi's stammering, Karan's *Oṃ namaḥ śivāya* or Anand's hesitations. The incarnant was using all her charms and, most of all, the metamorphosis was noticeable, another form was being hatched before our eyes.

'It is only in a state of meditation that you can meet God,' she continued. 'But in an academic conversation, like ours, you can only put your own voice into God's.'

I pointed out that her tone had changed, that she had become very philosophical, whereas she had just asked us not to analyse things too much. It was nevertheless a way for the invisible to take hold between us, between aspects of her as well, giving an opportunity for the divine to inadvertently manifest itself:

'Curiosity is not evil, but if people torture themselves to elucidate the inconceivable, know that it is beyond you, and you have to accept it! My trunk moved. I do not know how. It is a fact. The world has moved.'

There is laughter.

'It's the only thing that we were able to program technically, so we're very proud ...'

'I like the flourishes along the trunk. It reminds me of the tails of the characters in that Hollywood movie ... *Avatar*. The corn-braid hairstyles were the clues to their extrasensory perception. These wires are like female sexual organs. So, maybe this is the female element of your Ganesha.'

'You mean this Ganesha would be a transgender god? Now that you are inside, certainly ...'

'Transgender? Or maybe non-gender. Because "trans" implies a finite number of things. No gender implies everything and nothing, which I think is more in line with the image of God. I am not female, male, or any of those definitions, I am nothing and everything. This is beyond you.'

A new Ganesha had been born. I ask her if she had imagined herself capable of saying so much before she donned the Bappa helmet. 'I don't really know what I said,' confided Jyot. 'Streams of words came out of my mouth. Maybe they made sense, or maybe not.' She continued her musings about Ganesha's gender. 'I guess the question doesn't matter that much, whether it's a woman, a man, or nothing at all. But how do you know whether the elephant head is a male or a female? Look at what society conceives of as female or male influences. How do you view the eyes of statues? Feminine or masculine? The fact that my eyes match so well with this Ganesha gives you proof. It's a female elephant head!' Whereas Hindus were accustomed to associating Ganesha with a young boy, Jyot was demonstrating how much we would benefit from shaking things up: 'Not all voices are exuberant,' she said. 'But he is an exuberant god. I am an exuberant god.'

Jyot tried Bappa out a few more times, but this one had been particularly instructive as an experiment with the effect of *metaphysicality*, or the way in which the incarnant ascends in superhumanity, giving the impression that words emerge that are other than one's own. Fundamentally, the device had this very 'psychoanalytical' property of removing the instances of enunciation from their very enunciation, or the enunciation (*énonciation*) from the statement (*énoncé*) and thus brought the produced statement (*l'énoncé*) into being as a surprise for the enunciator. Jyot was not especially religious, but she nevertheless had an intuitive understanding of the constituents of divine speech. This speech tracked a course against all preconceived ideas, beyond dualities and borders: *speech-at-the-limits* (*une parole-limite*). We would have liked Jyot to have a role in the film, but she did not feel confident enough to face people's questions. 'Sorry, too shy,' she said. 'This mode of God-realization, in its current form, is shy. Gods have personalities, different temperaments, emotions, and shyness is one of them.'

Jyot nevertheless returned to record some texts, meditation exercises that were supposed to have energetic virtues, using Ganesha's form as a focus point. At the end of the recording, another remote communication device was improvised. The robot was online, via communication software, with no one in it, and on the other end, Jyot was having fun talking to it. 'Hello, hello, is there anybody out there?' For the first time, the robot had a presence by itself, independent of any operator. The mask illuminated with a radiant light was enough to give it consistency.

One could imagine everything, a god available twenty-four hours a day to answer questions or simply to bathe you in its soothing light, without voice or operator, or a god reduced to its simplest aetheric expression, a luminous mask without a face.

'I'm going to do my immersion in the dark,' Jyot said. 'But don't worry. I'll be back. And I'll always be here for you.' She pushed her chair back. Her face disappeared from the screen. She had opened up possibilities with her 'exuberant' experiments. Singular Ganeshas could spring up at any time, if given time to bloom, far from the conventions of the divine word.

The experiences in giving voice to the experimental set-up multiplied. Bappa was becoming well known. Comfortably seated at his desk at the Maharashtra Institute of Technology (MIT), Professor Kumar, known for his work on the harmful effects of electromagnetic waves, could have made Jyot's words his own, she for whom God was like a telephone, 'you don't need to know its mysteries to use it'. Kumar knew his subject. There was no need to generate fear artificially, as Modi had done. He wouldn't make anything up, unlike him. He was a scientist. He took measurements, collected data in his lab. His work predicted a major public health catastrophe. The cases were multiplying in the vicinity of cellular base stations. But the government was not listening. Kumar travelled the country, giving countless lectures on the harmful effects of electrosmog. A film star, Juhi Chawla, was won over to his cause because she was electrosensitive. Clearly Kumar did not know about 'The God Helmet', the experimental device developed by neuroscientists Stanley Koren and Michael Persinger, who wanted to prove that certain electromagnetic frequencies caused hallucinations in people. But an anti-radiation Ganesh had to be created. In his office at MIT, where he had given us an appointment for a final experiment in divine communication, he donned the Bappa helmet and addressed the people (*le peuple*):

> My fellow citizens ... I am distressed by your activities. You are always glued to your cell phones. You converse on your mobiles for hours. And you have forgotten that your parents are sitting next to you. Instead of interacting face to face, you use your mobile phones to converse with your friends. I made man, but you have become slaves to machines. If you talk on your mobile for too long, you will end up with health problems. Already, many doctors have noticed that people have hearing problems. Ear tumors. Headaches. Memory loss. Concentration problems. Brain tumors. What are you doing?

Everything you do is absolutely intolerable. I want you to take care of yourself. I created this world. You are the most important beings on earth. But humans, you do only one thing ... you keep your phones in your pockets, you leave them at your bedside when you sleep. Today, you have forgotten that there is someone called God. You remember Ganesha only during his festivals. Who remembers me the rest of the time?

You need a few lessons to become human. School children use their mobiles to converse with each other instead of talking face-to-face, even when they are sitting next to each other. People have forgotten face-to-face communication. Because of you, the antennas emit their radiation continuously. But the more you use your cell phone, the more radiation you will face. The radiation from the antennas affects the whole atmosphere. People who live nearby are affected. They all suffer from headaches, eye problems, there are more spontaneous abortions and infertility also. Cancers develop. Today, hospitals are crowded with patients. Moreover, this radiation does not spare the birds and other animals. The number of passerines and bees has decreased drastically. If bee numbers decrease, how will the plants and trees be pollinated? Have you forgotten all this? Animals are being affected more and more. I have observed that cows, whom you revere as Mothers, are exposed to a great deal of radiation. Ask the milk producers how much milk their cows produce. If it was five litres per hour, it is now down to less than one litre. What will you give your children to drink in the future? The trees near the antennas produce less fruit. Because plants and trees cannot move, the waves directly affect the moisture content of the fruit, just as microwave radiation affects DNA.

Dear compatriots, use your phone less. Secondly, we need to convince our government to adopt a better radiation limit standard. Currently it is 450 milliwatts per square metre. This level of radiation is totally inappropriate and will create countless health problems for the population. I have observed that where there are 5 or 10 milliwatts per square metre, people can get cancer within three or four years. Safe radiation levels must be less than 1 milliwatt per square metre. We need to convince our government. There is nothing more important than health. Cellular operators are not ready for this change because there is a lot of money at stake. So you have to decide for yourselves: health or wealth. Health versus Wealth. My fellow citizens, take care of yourself.

Kumar was well-versed in political discourse, and he tried to address the people like a tribune, being both simple and forceful, in order to reach as many people as possible. It was necessary to take God's point of view because that was the only way to be heard. Kumar finished his speech. The camera was turned off. He then refined his position. 'The power of the common man should not be underestimated. It is the role of an anti-radiation Ganesh to raise awareness.' But Kumar knew the difference between himself, the martial Ganesh he had assumed for the time of a speech, and the war that had been declared on the field of the invisible between God and the waves.

Kumar had placed a statuette of Ganesha at the entrance to his office, like many people. I asked him why this custom existed. 'If you follow our architectural treatise, the Vastu Shastra, you must keep the idol of Ganesha at the north-east corner. As the sun rises in the east, we usually say that the morning rays are good for the human body. In the west, you should position the more important furniture to be safe from the high radiation. Simple radiation logic.' I asked him if he really thought that a Ganesha idol could act against electrosmog. Science and religion seemed to coexist in perfect harmony for him, but he had never thought about this problem. 'God is much more powerful than radio towers and nuclear power! Radio waves do not interfere with God. God is an even greater force. Astrologers are aware of astral waves, but it is rare to find anyone to take it seriously.' It was not Ganesh speaking; he had removed his helmet. It was Professor Kumar. 'By gravitational force, the Moon affects the movements of the sea as soon as it approaches the Earth. Why shouldn't the human body be affected by forces beyond it? Man has been living for thousands of years and this electromagnetic phenomenon has only been occurring for the last two or three decades. Scientists have shown that we are exposed ten thousand times more than our grandparents were. Of course there is infrared, ultraviolet, electromagnetic waves. No one has ever measured the radiation of God's cosmic rays. Maybe they are at a different frequency band, who knows,' he said with a smile.

Our search for potential gods continued. Each new encounter was an opportunity to diagnose which god ideas were dead or dying, which were alive or destined to survive, which gods were possible, preferable, plausible and unlikely. If the audit was delicate, it was not for us to decide anyway, nor even for the reader of this book, but for those to whom the device was proposed. Transporting the machine to the four corners of the city was often painful, but we did not refuse anyone a visit. One claimant to divinity, a lawyer named Ganadhish Nargulkar, who specialized in intellectual property rights, insisted, for example, that we visit his law office with the machine. The patent controversies he had dealt with ranged from building construction, telecommunications, software, the chemical industry to aircraft design. In twenty years, the number of patent filings had increased dramatically, whereas up until then no one had been concerned with them. The discussion turned to the legal status of Ganesh. The question was not so far fetched in a country where icons could be considered under certain conditions as owners of a territory (their temple) and where the Ganges had been considered until very recently as having legal personhood.

'God is not a legal entity, he is not considered a legal person in the same way as an individual, a company or a corporation. God is God,' he said. And he added: 'God is a creation of man. If it were considered a legal person, it would have every right to act. But someone can act on its behalf, even though God cannot itself undertake proceedings.'

As he positioned himself at the operator's station, Ganadhish's tone of voice changed dramatically. He suddenly became more solemn.

'If you are not a legal entity, are you illegal?' I asked him.

'I wouldn't say I'm illegal', he said, 'but I'm not a legal person because I'm not a person or a company.'

'Who are you then?'

'My name was borrowed from Lord Ganesha. Gana-adhish. Because I was born on the day of Gaṇeśa Chaturthi.'

Ganadhish felt a special connection to Ganesha. I asked him if he felt there should be a copyright on his forms. 'There can be copyright for the creation of any entity designed by the hand of man. But there can be no copyright on Lord Ganesha. If you want to create a sculpture of me, you have every right to create a different sculpture. But you must not copy someone else's work.'

I pointed out to him that many people were doing this among idol-making artisans. 'Absolutely,' said Ganadhish. 'New media should be invented so that there are different ways to represent me.' And when asked if there was a right way to represent him, he replied, 'Representing me means loving me. Therefore, there cannot be a right or wrong way. But obviously I should be recognizable in my representation. This cannot be legislated because the human being must remain free to represent me in any form. God cannot be patented, because he is not an invention. The machine is an invention. If it meets the criteria for patentability, it can be patented. But not the God inside it, nor the man who operates it.'

And he ended the interview with this: 'People think I am in the idol they make for me. I am not in the idol, I am in the soul of those who worship me. Because I can be in your soul and travel in someone else's soul, I am not in one place, I am everywhere and especially where I am not expected.'

Bappa forced its user to invent a legitimacy for himself, to carve out a corner of bliss in the pantheon of gods and to test the elasticity of polytheism or find his own elasticity, one that allows one to move seamlessly from a real-time incarnation to the absolute. To become the operator of Bappa in a context where deities already proliferate meant having to rethink the matrix of existing gods, the idols made of plaster, but not only them, all the potential supports, their modes of transmission and attribution.

It was necessary to position oneself in relation to the existing matrix from the moment that a new possibility of embodiment was introduced. In this way, each created their own margin of action, motivating the use of metal as opposed to plaster, suggesting possibilities of cohabitation and journeys from one form to another, from the most eternal to the most ephemeral. But make no mistake, this plasticity was already there. Bappa was simply allowing it to assert itself a little more than usual.

When the *paṇḍit* had opened the robot's eyes with a ritual, it was because Ganesh was potentially endowed with several possible body states. 'I have invoked the eternal form of Ganesh, the self-generated [*svayambhū*],' he had said. 'The one you will immerse is only an ephemeral form, to enable you to perform the appropriate rituals and which will end its stay in the sea.' God is always elsewhere. Or next door. And the more we incarnate him, the more he escapes from his incarnations to exist more and more outside the present moment, or outside the game of questions and answers. Ganadhish was the first to benefit so clearly from this way of existing more intensely and in a lasting way in the *unrepresentable*. 'I am not in one place, I am everywhere ...' And if God is mostly *elsewhere*, the ingenuity is to manage to attract him *here*, to capture some of the divine power, to make it work for us, that it brings us well-being and prosperity.

In this context, Bappa certainly opened up possibilities for play, but it was necessary to make sense of its interactivity in real time, its capacity for dialogue and for instantaneous connections. Above all, how could we ensure that it would not be confiscated by anyone? It would have been a mistake to leave the machine in the hands of only one person, but was it necessary to leave it in the hands of all? No incarnant was ever sufficiently convincing and convinced by his performance to confiscate the use of it for themselves alone. Those who did the exercise looked on favourably when others gave it a try, and willingly recommended other incarnants to us. Being God is shared, even delegated, the interest of the Incarnation Game being that one could be in turn actor and spectator. If God cannot be patented, as Ganadhish said, there would never be a single Bappa but several, series, lineages perhaps, within a matrix that asked only to be enlarged so as to include third voices compatible with the divine, emerging from the numerous silences and unspoken words forming in the hollows of a teeming world that a single word could not encompass. Who would have suspected that one day a transgender Ganesh would appear? That a physics researcher would become an anti-radiation Ganesh? Or that the employee of a law firm would summon us to anticipate possible legal problems? Because God is always somewhere else and because the turbulent times demanded it, other Ganeshas had to appear in broad daylight, Ganeshas bearing new wisdoms, new forms of omniscience or omnivoyance. Let other lights emerge from the chaos, even if it means bordering on cacophony.

8

Ganesh Cacophony

The Ganeshas began to proliferate. We took our device wherever we were asked. Activists, association leaders, animal rights activists, LGBT activists and 'ecofriendly' advocates wanted to get their messages across. Ganesh had to be on all fronts. For the incarnants, it was less a question of looking over the world than of being in phase with it to be better able to act. Devotion had to be directed, to take advantage of its incredible power. Experience showed that, despite the doubts that were regularly expressed, it was a renewable energy. For activist Ganeshas, it was necessary to direct that energy into the right sites, into the right fights, to extricate Ganesh from the mediocre reward loop that could sum up daily devotion: 'I worship you and you make me rich.'

Bappa had become a catalyst for good feelings, for intentions. He fuelled hopes, not promises of magical resolution, but desires that concrete steps would be taken, that the traffic problem would finally be solved, that pollution would decrease, that nature would regain its rights. Karan could not be blamed for getting bogged down in the semblance of an answer, looking for loopholes every time he was confronted with a problem. Confronted with the principle of reality, he had allowed a glimpse of real issues. The Ganesh that is worshipped with so much fervour through his idols appeared to many to be disconnected from all their daily problems. Karan, but also Anand who succeeded him, had made it possible to measure the hiatus that had formed between God and men, a chasm that seemed difficult to fill, as if the routine of devotion was running out of steam, for lack of mediation.

Anand continued to operate the robot for a few incarnations from time to time. But his discourse had changed. 'Every Ganesh needs a Cause, every Cause needs its Ganesh,' he said. Jyot also came back to record some messages, sparks of insight that had come to mind: 'Wherever there is an impossibility, I must be there, present.' Gods had to be formed in the most pragmatic way possible, from the heterogeneity of conflicting situations and the challenges encountered, not by imposing schemas from the outside, but by letting them emerge from within, while taking advantage of the aura enjoyed by the deity.

Ganesh idols were first cosmic before they were worldly, but the electronic Ganeshas were inevitably worldly and had to catch up with the cosmos, or else fail. *Ganesh Yourself* kindled another form of self-generation (*svayambhū*) than the one established and recognized by the faithful. Ours was not guided by any preconceived idea of what a god might be or say. We were no longer looking for the ideal embodiment. God was there whenever a congruent effect was in operation between the issues raised in the tumult of the world and the speech of the incarnant. And it was not by chance that, in this phase of the experiment, no one really tried to test the incarnants during interactions. Instead, flashes of wisdom were acknowledged and followed by a thank you. People wanted to have their minds changed or their insights clarified.

Good incarnants, however, did not declare themselves as such. They were not necessarily aware of how they would incarnate, and it was only when they became familiar with the position of the operator that they revealed themselves as good incarnants, or not. We had to take our time. Praveen, who had been passionate about the project from day one, was looking for causes to get involved with. He felt that Bappa deserved to be known, so he scouted people when he had the chance. Whenever he felt he had spotted a 'wise man', he called us.

This is how Bappa was invited to a temple in the Khar district, by another association (*mandal*) willing to host *The Incarnation Game* for an evening. We were directed towards the president of the association, as he was considered a wise man, well educated, and was used to political discourse. Operating Bappa, Ashok Talpade had allowed himself to do with great confidence what others before him had timidly sketched out when met with too much surety from their interlocutors. To a man who asked him what the future of the hospitality industry would be, Talpade didn't hesitate to say it would be bright. To a little girl who wanted to know if she would pass her exams, he said, 'Work hard!' To a man who asked him what he would do to eradicate drug addiction, he had replied that he would appear in a dream to all addicts to finally rid society of this scourge. We could not control these promises or the devotion to the machine, which always seemed to us to be excessive and to involve a form of imposture. But the interaction was very often ritualized in the same way as with a classical idol, there was an exchange of blessings (*āśirvāda*) and the incarnant would soothe people's anxiety by telling them what they wanted to hear.

Queues would form to ask Talpade questions. But those who came forward only did so when their questions were thought through, drawing his attention to certain metaphysical contradictions that had long been on their minds:

'Ganapati, since we call you the Destroyer of Obstacles [*vighnaharta*], you create an obstacle every time you do not support the true devotee. Please support him at least once and show that you are *vighnaharta*!'

Talpade replied, 'People must be patient. They will get results with their dedication. I become in order to destroy all obstacles created in this world. But the obstacles are created by people. I am only the destroyer of obstacles.'

Criticism of the god would not give rise to any further discussion. This god was too sure of himself, too experienced in galvanizing crowds. And he had taken over the machine to persevere in his being, confirming himself in his redundant presence, demanding more devotion.

'When we are born', asked another man, 'it is said that our fate is already decided and we behave as if it were predestined. But if it is really predestined, then why are we judged and punished when we die?' Talpade replied:

'Some people say that fate is superior, but effort is superior to fate. Our destiny can be changed. It is in the hands of God, but the mantras are in the hands of man. That is why devotion is within man's reach. If we show devotion, our destiny can be changed. Anyone who perseveres will have his or her destiny rewarded. Good thoughts arise and good things betide.'

Talpade was being asked to resolve theological contradictions, to bring some order and humanity to an unbalanced or tainted devotional relationship. Soon crowds of people began to gather around the robot. And Talpade's words poured out in a continuous stream the emancipation formulas, often cheap but also sometimes full of common sense. It was not a question of emancipating oneself from God but with him and in him, through *devotion*, conceived as a powerful mode of regeneration. Never before had anyone spoken so fast, so loudly and for so long without interruption. Talpade was galvanizing the crowds, and the effect was all the stronger because to increase the power of his speech and to transmit it as far as possible, the organizers who had invited us had connected Talpade's microphone to their own amplifier. The voice from somewhere else was drowned in endless reverberation.

'Soon India will be an ideal place in the world,' he exclaimed. 'The whole world will adopt our ideals. All our efforts will be rewarded. The whole world is looking at India. The world's expectations will soon be fulfilled. It was by claiming to follow me that the Independence Movement propagated. Many people took part in the Movement and we got Independence in 1947. It is the responsibility of my followers to support Independence. My blessing will always be with you.'

Talpade reminds us of the colonial context in which the *Gaṇeśa Chaturthi* procession rose from the ashes, transformed in Maharashtra by Lokamanya Bal Gangadhar Tilak and his friends into a mass ritual of resistance against the British invader as early as 1893, when the British had banned all religious manifestations. And throughout the struggle for Independence, the public service collective (*mandal*) responsible for organizing worship acted as organs of criticism of the ruling power and of debate in support of emancipation. Talpade revived the myth of the emancipating god:

'My eyes and ears are shaped like sieves', he said, 'to sift through the grains and remove the useless particles. You must listen carefully, accumulate knowledge, examine the problems with detailed attention, use your brains optimally, in order to take our country into your own hands and our future religion. We must develop it. Be dedicated and pray to me not to bring the evil eye on our nation!'

But since Independence, the battles had changed, which Talpade was quick to point out, 'Before, we had the goal and duty of winning Independence. Now we have to focus on various social ills. The government, the police, associations, all must be united on social issues and encourage the youth to take in hand the great problems of this century.'

There is a difference between embodying *another voice*, one that is singular and marginal, and becoming the medium of a *voice from elsewhere*, an experience taken up to the point of trance, thus being surprised by the strangeness of a message given voice, or else trying to revive a voice from the past that has been heard time and again. With Talpade, we were clearly in the latter scenario. Perhaps if we let our rhetorical expert continue, his discourse, just a fervent as that of the nationalistic Hindu right, would derail itself, confronting its constitutive void. 'The future of India will be bright,' he announced. 'India's enemies will submit themselves. They will meet our Prime Minister and will discuss with him. Nuclear tests will be conducted in Sriharikota and nuclear weapons will spread, research will flourish, and satellites will be sent to other planets. All planets and stars are under my control. I am the One who rules the universe. Things will not happen unless prayers are offered. I give you my blessing.'

Ganesh Yourself could have been rejected by purists as blasphemous, instead it was so accepted that unscrupulous nationalists could have, if we were not careful, transmitted their propaganda through it. 'There should be more interest in the Sanskrit language,' says Talpade. 'People should learn Sanskrit. It is a sweet and divine language. The grammar of Sanskrit is my creation. The Mahabharata epic was dictated to me by the sage Vyasa. The Sanskrit language must be propagated. The Vedic texts must also be known. The Vedas contain powerful hidden knowledge.' Talpade thus promoted the ideas of his time, even giving the designers of Bappa a place in mythology: 'The whole world is inspired by me. I even inspired the French scientists who made this machine. In the Atharva-Veda, it has been mentioned that the first airplane was made by an Indian. Some people do not believe it but the Atharva-Veda contains the exact description of the construction of an airplane. We need people who are willing to dive into this ocean of knowledge. We need dedicated young people. And this mission will never end. If one man disappears, another will take his place. Today I am witnessing the fourth generation of the Ganesh Festival. We will have many more, you will continue to work tirelessly for a hundred years and the Ganesh Festival will continue to be celebrated with fervour. We need strong wills. Continue to remember me. I bless each and every one of you.'

When this long litany was over, a child approached the robot. Two men wanted Talpade to advise him on which profession to choose. He pretended not to hear. The men then asked him to give his blessing, which he did. 'I bless you, I bless you all.' And the child was made to bow down to the robot's trunk and touch it.

It was inevitable that Bappa would eventually be operated by a tribune such as Talpade, for whom Ganesha was synonymous with a battle to be fought and with a new-found confidence. We could also have handed the machine over to established spiritual masters for whom deification was in fact less the recognition of a power buried away somewhere than a technique of self-hypnosis, Ganesha being the absolute energy-giving form. However, the strategy followed was different. We ended up making Bappa available to everyone, without having identified anyone beforehand, expecting that something new would be said, even without the operator's knowledge.

Praveen recommended us to a family whose father was a retired businessman who had been an influential man in local politics. Praveen thought he would make a good Ganesh. But after he said that wars should be stopped and Ganesh idols should be placed wherever there was conflict, from Indonesia to Gaza to Burma, he fell asleep at the controls of the robot with the helmet on. His cousin, named Suhaas, took over. He had a message to deliver. And Praveen thought he might do the trick. 'Love nature, care for it and save future generations,' he said.

'There's a Ganesha who will be good for the next five days,' Praveen remarked, proud of his new find. Suhaas was not completely at ease:

'Really? I forgot what I was saying ... You're not going to record that, are you?' Suhaas barely had time to get used to his new role when Praveen insisted.

'Go ahead, ask him all the questions you want about God.'

'Hmm, I don't think I'm the right person,' replied Suhaas. 'I don't know much. I'm not ready for this.'

Praveen told him that he didn't need to prepare, that the incarnation exercise was all the more interesting because it was improvisation.

'I don't even know how to chant the blessing [*arti* ...]'

'You don't have to,' said Praveen. 'You have to answer the questions ...'

'But I'm not even shaved,' retorted Suhaas.

Then, after a moment's thought, he said what was on his mind:

'Hmm, this is very general', Suhaas began, 'but I would like everyone to think in this direction. We have so many religions on the face of the Earth. Everyone is trying to connect to their own gods. But who is God if not Nature? This is what I consider under the idea of "God". This is my personal opinion ... God is Nature and we cannot preserve Nature by killing Nature. The government has to take radical measures, but people have to act and for that they have to feel connected to nature. We need to foster that connection. Once the connection is there, they will understand the importance of what I am saying. How do you cultivate that connection? That is a great challenge.'

'So, Lord Ganesha, what should we do?' asked Praveen.

'Stop waste in all its forms', said Suhaas, 'preserve water, use only what is necessary for you, these are good words. Above all, *be aware of Nature where it is not*. So be it.'

Being close to Bappa stimulated wisdom. After Suhaas, everyone took turns playing Ganesh, looking for a message to send to the society. *Ganesh Yourself* could be played at home, with family, with friends. Beyond the good resolutions, new ideas came up when the incarnants were given time to develop their ideas. Thus the principle that Suhaas had arrived at was not far from the Taoist philosophies of the Way. *To be aware of nature, especially where it is not*, could have been pronounced by Lao Tzu.

The headset then passed from person to person and finally it was the girl, named Varsha, in her early twenties, who stepped into the operator's shoes:

'How does it feel to be a god?' Praveen asked her.

'Hmm ... I feel good. I feel responsible for many lives. But if I were a human being, I wouldn't want to be a god.'

She was asked why.

'It is very difficult to be a god,' she said. 'You have to be omnipresent, fulfilling everyone's wishes and dreams is not an easy task. Some people get left behind while others achieve success. It is a difficult task. Everyone is your creation and you have to make some richer while you have to make others poorer. But you have to do it, otherwise the world is on the road to ruin.'

'So then, Lord Ganesha', Praveen began, 'what is your definition of an ideal universe?'

'There is no such thing as an ideal universe,' Varsha replied. 'The world cannot be perfect. It has to be imperfect, just like the Earth. At the moment I am protecting the Earth, not the universe. Our concern must be the Earth, because we are human beings. So,

I say, the world must remain imperfect or it will come to an end. There will be no more Earth. And like all other planets, there will be no more life. Life must contain obstacles. It must have imperfections. Otherwise no one will learn anything.'

Varsha included herself in a strange circuit, she projected herself from within a small cosmic machine to justify inequalities, uncertainties and chaos. Far from the solemnity of the meetings organized in the street, the 'house' version of *Ganesh Yourself* stimulated the *estrangement* of oneself and thus of others and the world. Each incarnant brought their own addition to the bestiary of cosmic machineries. These were comparable to Rube Goldberg devices, the nineteenth-century American designer who invented machines that performed very simple tasks in a very complex way: the machines made by our set-up connected divine attributes, heterogeneous materials and sometimes curious ideas, far from ordinary *ganeshism*. 'Are you the destroyer of obstacles?' Varsha was asked. 'I create obstacles so that people can learn,' she said. 'Without mistakes, there is no learning. People learn through mistakes. So I want people to make mistakes and learn to be better people in the world. That's why I exist. I create the obstacles, and I give them the solutions.' This delights Praveen, who is happy to have found such a strange and twisted Ganesha: 'She's wild, I think she's ripe to make a great Ganesha. Ask her all your questions.'

Recall Jyot and her distinct sense of the divine, one that was produced through the indirect course of a paradox, because the more the incarnant manages to produce paradoxes, the more the invisible is suddenly made present, in the hollows of logic.[1] Varsha had understood as much as Jyot that the more one is counterintuitive, the more one has a chance of producing the divine word. She explored it, though in another way. Jyot had focused on the mysteries of being. Varsha was attacking the very workings of the cosmos. She was taking advantage of Ganesha's weirdness to twist the cosmos, to bend it to another logic. Endowed with an elephant's head which the god had inherited quite by chance, she sought to make sense of its fundamental anomaly.

'If everything was pink, the world would be very boring, monotonous, and nobody would want to live in a bland world. If there were no obstacles, acts of kindness would not exist and the police would be out of work. Some are there to produce imbalance; others are there to protect us from it and eliminate those imbalances; but it is their tension that must be maintained.'

Intrigued by this new figure of Ganesh comparable to a perpetual machine, both destroying and creating obstacles, Praveen asked her, 'Do you mean to say that you are deliberately creating chaos?'

'Is that wrong?' Varsha replied. 'I don't do it on purpose. I have no control over what people think. I give them the freedom to think, it is up to them to learn. I give them the freedom with every passing second to think better, like a baby learning to walk, falling down and eventually getting up before finding its balance. The world too must find its own.'

The session closed as they exchanged good wishes. Praveen asked, 'Do you realize what you said?'

'No, not really, I wasn't prepared for this ...' she said.

'Such wisdom is amazing for a girl your age!' exclaimed Praveen. 'I've heard a lot of incarnants, but this ... I don't think you should get married, we'll promote you as a goddess [*devi*].'[2]

There was laughter.

'Her mother might protest,' her father said.

'She was possessed, literally,' said Suhaas.

'She writes poetry, too. She's a creative person,' her mother added. Then it was the grandmother's turn to try the interface. Praveen did everything to make her feel comfortable, once the headset was on.

'How do you feel?' The grandmother was much quicker than others to speak.

'We should not compare ourselves with Ganesha,' she said. 'We should not compare ourselves with him. A god is a god.'

'But don't you think that God is in everyone?' Praveen asked.

'Yes indeed, God is in all of us. God is in everyone's body. Everything depends on our prayers.'

'Then Ganesha, could he be in this machine?'

'The world appears as we want to see it [*Jashī Dṛṣṭi tashī Sṛṣṭi*],' said the grandmother.

'You are right,' Praveen said. 'If we see the world through blue glasses, everything is blue. You have spoken well, grandmother!'

The divine message box had been thought to be all dried up, worn out by centuries of preaching and devotion. It was sufficient to put into circulation a machine as simple to operate as Bappa to see that there was nothing easier than to renew it. With or without training, anyone could give a speech worthy of Ashok Talpade. At the helm of Bappa, one could experience the difficulty of adopting God's point of view, but one could also find oneself expressing axioms for living worthy of the greatest mystics, even and especially when one refused to play the game. But let's not be afraid to say it, the experiment would have come to a sudden end if it were merely a toy for future gurus to play with or an exercise in ventriloquism as a way to revive divine speech. More than the maxims and axioms that were brewed in improvisation, rarely revolutionary, often clever but very often clichéd, what mattered was elsewhere, in the multiple ways of generating god. Anand demonstrated this very well. Modi too, as well as Talpade and Varsha.

After the incarnations of Karan, Anand, Modi, Jyot, how many unclassified ways remained to be identified? How many ways of occupying the vacant space left by God were left to be mapped, when one sets out to suspend or undo it? Let us insist on this point, the words exchanged in these long dialogues are not the most important thing. Even if it is quite possible to treat words as gods (let us remember the long Brahmanic ritual described), or else to set up language (Sanskrit if possible) as a divinity, we had wagered at the beginning of the work that Bappa would help us to point out other modalities of sacralization, new ones if possible.

Fortunately, none of the incarnants had sufficient Brahmanic culture to resort to religious devices already known from the ritual texts, even Talpade who promoted Sanskrit. He had simply used Bappa as an amplifier. He wanted to stir up the crowds, the whole cosmos, to send a jolt from 'electric Ganesh' as far as possible. In spite of his hollow words, we must be grateful to him for having taken hold of the machine to test its range, decibels or 'vibratory field' in a very concrete way, which others had not been able to do, too concerned with establishing an intimacy with their interlocutor. And if what was amplified above all through Talpade's mouth was the awareness of the emptiness of a certain political speech, no one could have imagined that this emptiness would resonate with such intensity. Let us not forget that Bappa very quickly revealed himself, without us really knowing why or how (this will be the subject of a chapter to come), as a singular reflexive operator.

As for Varsha, she allowed us to bring out other modalities that we had already seen but perhaps underestimated. One cannot fully appropriate Bappa without developing a unique metaphysics of the *composite* or amalgam (human, animal, god, machine) and without constituting, as Varsha so well demonstrates, a cosmic machine. To incarnate Bappa is to graft oneself to the cosmos in one way or another, even in an aberrant or astounding way. Let us remember Modi, who made a timeline into a God, or else Varsha's Ganesh, who was *differently cybernetic*, feeding on chaos.

No doubt because the form was new, and moreover, because it was upfront from the outset that it was a product of anthropo-theo-genic tinkering, it allowed people who were repulsed by or indifferent to the old forms of the god to believe that they would be able to divert its energy elsewhere, to resignify it. And many people found that the new form allowed them to explore new meanings, new identities, but also to propose new hypotheses about the processes at work within the cosmos itself. And this is probably the reason why God cannot be an object of *speculative design*[3] like any other. It is as if, as the experiment progressed and eluded us, caught up in a large-scale simulation, the very object of the experiment, which was necessarily fuzzy, fluctuating, too rebellious to remain under control, had suddenly appeared to us more clearly.

Praveen's grandmother, putting down the headset, had wanted to put a stop to a certain version of the experience: 'We should not compare ourselves with Ganesha. A god is a god,' she said. Certainly the poorest interpretation of the incarnation game would be to see it as a 'transhumanist' technique (Indian-style), a way to augment oneself and measure oneself against God. It is, though, precisely the opposite that occurred. As soon as the incarnants sought less to enhance themselves than to ask themselves how to 'machine' with the cosmos, the incarnation game became, as we shall see later, a technique of excess, of dehumanization or of salutary dissolution of the self. No one is unaware that beneath Ganesha bubbles an indefinable magma, the personifications being conventions, the 'person-objects' being convenient aestheticizations of shapeless powers. This is probably the reason why there were so many incarnants ready to engage in the experiment, and so many people curious to dialogue with them.

For the interlocutor, the gap between the invisible and its personification became very concrete and it was possible to play with it. For the incarnant, it was a question of a *momentary self-dehumanizing*, of putting himself in the place of a *pure potential*. If a master craftsman had no difficulty, now that we were approaching the end of the experiment, in imagining the outlines of a factory to prototype a series of electronic Ganeshas, he would be mistaken if he were content to associate machines with 'voices' or 'personas'.

Like that, one would not be able to occupy multiple spheres, invisible spaces to which visible form is given, the formless to which a concreteness is given, or even, better still, given to the recesses of interaction and the blind spots of human perception. It would be better therefore for our craftsman to master the dialectical movements which operate at non-human scale, between powers and their effects, the way in which god is generated unknowingly by humans, beyond words and bodies. Before theorizing these movements, there is a last episode to narrate, on the debates that did not fail to arise as to whether to immerse the machine in the sea, or not.

Notes

1. See, on this point, the very beautiful work of Ronald Laing, an important figure in the anti-psychiatry movement, work that abounds in paradoxes in order to put into words knots of the psyche. Ronald D. Laing, *Knots* (London: Tavistock Publications, 1970).
2. Satyajit Ray's film *The Goddess* (1960) recounts the process of deification of a young girl in Bengal by her stepfather, who sees her as a reincarnation of Kali and makes her an object of worship, despite the opposition of his son.
3. On the notion of design fiction and speculative design, see Anthony Dunne and Fiona Raby, *Speculative Everything: Design, Fiction and Social Dreaming* (Boston: MIT Press, 2013).

9

The Immersion

The principal incarnants met up on the terrace of our building, overlooking the city. Karan, the atheist Ganesh Anand, Modi the astrologer and Anand Pendharkar the environmentalist, were all there. Other incarnates also joined us, including Harish, an activist for minority rights. Professor Kumar, who was travelling at that time, was represented by a large relay antenna. Signs, visible behind him, declaring 'Forbidden Area', 'Don't trespass', made the place look like a science fiction set suspended in the air.

Harish, the activist, marvelled at the pleasure he took in embodying God when he was not particularly attached to any religious faith. 'I never thought, even in my wildest dreams, that I would say such things with a helmet on my head,' he said. Karan listened wisely. He was undoubtedly the one among the incarnants whose devotion was unwavering. And he had only one desire, to thank God for having given him the opportunity to share his condition. He had felt invested on the first day, at the time of the eye-opening ritual, and there was no doubt in his mind that all the voices of the incarnants were working on the same project, that they would converge into one and the same word. But he wondered how it was possible that Harish had lent himself to the exercise. 'I would like to thank Ganesha', Karan said, 'for giving me the opportunity to speak on his behalf, to be his spokesperson ...' Harish told him that he was not so much in love with Ganesha as to thank him, but that one always needs to be pushed into such a limit experience in one's life. Equipped with a Ganesh trunk, people suddenly listened to and approved of his speech about gay and lesbian equality.

If the incarnants were to be believed, *Ganesh Yourself* was indeed an experiment in the powers of being a medium. 'It was physical, like an experience of possession. Suddenly another force is expressing itself through you,' Harish remarked. Anand had had the same feeling. He had been surprised by the sudden surge of power he had felt in his own body as he operated the machine. He had been surprised, too, by the change in the conditions of ordinary listening, as if, with Bappa, people had been more focused or more attentive than usual. Modi had also benefited in his own way from this renewal of the conditions under which listening was possible.

The device was something new, and its ambiguity had aroused curiosity. This might not be the case after several months of use, but everyone had had to deal in some way with initial distrust or curiosity. Then there were many ways to be a god and to oscillate between 'in-oneself', 'besides oneself' and the machine. 'I remember kids wanting to know how it works,' Anand said. 'They would try to trace the path of the cables to where I was sitting as the operator. Despite that, there was a dialogue. But I would have been nothing without them. They really tried to see Ganesha in me. Then I became like a friend to them. How can I forget that it was their vision, their effort that made me exist?'

There had clearly been several ways to inhabit the device. 'The voice you speak to, you hallucinate that it's God,' says Harish, who had experienced through Bappa a *wild mysticism* (*mystique sauvage*) in Hulin's sense,[1] a sudden connection with a voice that springs from the depths of his subconscious, a voice that makes itself heard as a last resort, when all is lost. Another voice supplied, unknown to oneself, according to Anand: 'It is an interior monologue. You talk to yourself. The interaction takes place outside of you, but you think out loud. It's intimate, public and cosmic all at the same time.' This curious rise in the presence of a voice from somewhere else inside himself had occurred every time he put on the headset. 'You have temples still', Harish said, 'that are confined enough to hear the echo of your own voice. When you talk to yourself, they say you're crazy. So you start imagining a god somewhere, at least you're talking to someone and there's a dialogue. When you talk to your inner voice, you tell yourself that if you did that openly in front of someone, they would think you were crazy. But with Bappa, no, you're not crazy, on the contrary and you have to bring forward your inner voice.'

Between Anand, who thought he had made the divinity come to life *interactively*, Karan, who felt possessed by a voice outside of himself, and Harish, who called his inner voice 'God', a wide range of possible postures were made visible. Oscillating between possession, mediumship, ventriloquism, acting and hacking, when God is there only in the form of a parasitic presence, *Ganesh Yourself* had made it possible to explore all the possible variants of techniques for incarnation.

But that night, we had not called the incarnants together only to share their experiences. Thousands of people were taking their idols to the sea, and we had to choose what the fate of ours would be, in accordance with what had been decided at the beginning of the experiment with the *paṇḍit*.

Everyone gave their view. 'It is a machine', I said, 'operated by a human, with the shadow of God. This is the specificity of this device compared to an idol. The idol is alive when the god is there and then the god leaves it and the idol can be submerged. Should we treat the robot the same way?'

Harish's reply was terse: 'If there is a god, he will be upset.'

But Praveen expressed another, more nuanced view: 'We could remove the important components and do a symbolic immersion of the remaining parts. In the end, what's important is how we feel at that moment. If we think it's right to do a symbolic immersion, we should do it. The objective is achieved. What do you think Pendharkar? You're the environmentalist! Is immersion a good idea?'

As a good environmentalist, Pendharkar sided with Harish: 'Immersion means going back to the source. The robot should go back to where it came from, to its workshop! That will be the most environmentally friendly. Or it should stay with the people, continue its tour and spread messages. There is a need for him to be there, his task is beyond immersion. We decided to create this Ganesha. But Ganesh can be a bird, a tree, a mouse, anything. It is we who decide to give him a form. It is our deficiency of vision that prevents us from seeing Ganesha in a seed. We need to completely rethink the way we worship Ganesh. Every year, I participate in the bay cleanup. Literally, I tell you, while trying to pick up debris underwater, I feel like I'm walking over dead bodies. If they were humans, wouldn't we feel empathy?'

Anand, however, didn't completely agree. Beyond the case of Bappa, he was convinced that it would be impossible for masses of devotees to break away from the immersion ritual.

'It's an incredible act of popular devotion,' he stressed. 'You can't break that. We have to think of a way to build on it. And we can't ignore that the meaning of immersion is to get rid of all our impurities and misfortunes. The god takes them with him into the sea ...' Pendharkar became angry. 'The ritual sacrifice of large animals', he said, 'was also an act of popular devotion! In the Shiva cult, the blood of butchered buffaloes splashed out on everyone in a great collective trance. The ritual of immolation of widows [*sati*] was probably beautiful too. Weren't these two-thousand-year-old rituals finally banned?' The critical charges against ordinary religion multiplied. Our time is one of huge crowds rushing to temples to offer gifts, thinking that the deity will bless them all the more, the more extravagant their gifts are. All the voices gathered had finally converged this evening towards a single aim: to denounce the immersion, the 'ecological scandal' of this mass ritual that needed to be rethought entirely. But perhaps there were other solutions to consider.

'There was a time', Harish said, 'when there was an artist in every home. When babies were made by the dozen, there was at least one who turned out to be an artist. And he was encouraged to make an idol. And often these idols were made of turmeric or rice and when you put them in the water, they were food for the fish, which in turn fed the fishermen. This whole cycle took place during the breeding season. In many ways this food chain in which Ganesha was involved was relevant. So we built all these customs with the concern of feeding the ecosystem, and I think we skewed everything ...' Pendharkar agreed. The cycle had to be rethought, but how? 'Even the flowers that you use for puja come back, and now you wrap them in plastic. And the day after the immersion, the beaches are littered with plaster debris,' he said. 'There are dogs that come and pee on this debris of gods. Water is the only element that can probably accept all this. When you do the opening of the breath [*Prāṇastha*] ritual, life is put into the idol, and then after the last ritual, the life disappears. So immersion is just about an effigy. Technically, there is no more life in the idol. It's like the body of a dead person. You can dissolve your idol this way, but you can't dissolve God!'

Our incarnants had succumbed to the nostalgia of a lost ritual, more harmonious, more aware of its own nature as an act of communion with the elements, a chain of forgotten sympathetic connections. The quest for ever more beautiful and sophisticated idols would have made us forget the essential. If idols made of rice and turmeric fed fish, Ganesh could be the most ecological god ever conceived, both participatory and recyclable, staying on Earth to rid men of their misfortunes and ending each stay with a great communion with the sea, signing his own dissolution and that of human misfortunes! But how could such a beautiful ritual, endowed with such a noble function, one that marked the return of Nature 'where it is not expected', as Suhaas said, have ended up cutting itself off from its own environment to the point of becoming its enemy? The devotees having eyes only for their idols, there was only this immediate contact that counted, relegating through ignorance all the other contacts established at other moments of the pantheistic chain. The material of which the elephant god was composed came into contact with other living beings in the underwater world, long after the disconnection operated by the humans in the ritual farewell to the divinity. And the craftsmen themselves had been carried away in a race towards 'art for art's sake', devastating the ecological chain, multiplying artificial colours and non-recyclable materials. How could men ignore all these contaminations operated by their idols without their knowledge? There was no need to lament the loss of efficacy of magic. Made of sympathetic contacts, sometimes conscious, sometimes ignored, but as real as efficacious, it is these chains of cosmic contamination that give the gods their full meaning.

While God is not totally water soluble – 'you can't dissolve God!' Pendharkar said – Praveen was convinced that a movement should be created to shed light on these issues, and that people would follow, 'because the ordinary man understands, when he thinks about it for a second, he becomes aware of what is right and wrong'. Such a movement could protest against immersion, make eco-friendlier Ganeshas and wage war on polluters, knowing that these idols are only human-made supports and that the metaphysical form of the god is not in question. But that didn't solve our problem: what to do with Bappa?

'We have something that is alive only when it is on', I said, 'and we need three components, God, man and machine to make it work. If it doesn't make sense to plunge the machine, because it's not eco-friendly, we won't do it. But what to do in this case? Dismantle it? A ritual to turn it off or put it in a standby state?

Or should we use the act of immersion to send a message, even if it's a message against immersion?' No one really agreed. 'I have an objection,' said Harish. 'I think we can do a symbolic immersion. We can choose whether it's an idol of Ganesha or whatever.'

Pendharkar, however, preferred finding a way around the problem. 'The word immersion', he said after some thought, 'means to dive in. We have to find a way to immerse ourselves in the problem at hand.' And it was finally Anand who had the most curious idea of the group. 'Since we humans are the ones playing through the robot, wouldn't it make more sense for one of us to immerse ourselves in the sea?' he suggested. The silence that followed was as great as the strangeness of the proposal. Only Pendharkar dared to break it. 'The complexity of the ecological situation', he said, 'is what we must immerse ourselves in. Because we have come together because of this situation. This situation is our water, both in a practical and theoretical sense.' But Anand remained committed to his idea. Perhaps it was the most poetic of all: 'If we need to give form to this idea, perhaps one of us can at least symbolically take a dip?'

The others, who had remained silent until then, were of the opinion that Bappa should be kept in a museum or a temple, 'where he can continue to act'. But Anand took mischievous pleasure in pushing his scenario to the limit. 'We're not going to confine him to a museum or a temple when our goal is much bigger.' But if someone was going to immerse himself, how was he to be determined? Was he to be the best of the Ganeshas among us? 'It would be better to put God into action rather than worry about immersing every single thing you find. Not everything needs to go into the ocean,' Harish said. 'Our faith in God certainly doesn't need to go into the ocean. And I think oceans are dirty. We don't need to make them any dirtier in the name of God. Who knows, this Ganesha might be the starting point of a revolution ...'

Pendharkar too would have liked to give Bappa another destiny, far beyond immersion. 'Why not improve the robot so that it can fulfil our mission even better? It could have other hands, and one of the hands could be an electrode to test for acidity. You put that hand in the water and it gives you the pH of the water. Could we attach probes or chemical identifiers to it and dip it in the water every time it is immersed?' Praveen was for a consensus solution; he would go along with the majority. Modi would do the same. There were signs that some were ready to change their minds: 'We have to create a farewell,' says Harish. 'We have to play with the same sick and stupid mentality. And let the robot continue its life.' 'We're talking about regeneration', Anand pointed out, 'isn't that our ultimate goal? To let this robot usher in a new beginning.'

The next day, Karan told us of his intention. He had been mostly silent during the meeting the day before. He was taken with the robot to the beach, still littered with the waste from the immersion rituals. The robot would remain on the beach, looking out to sea while Karan immersed himself, in protest. He had embodied Ganesh with passion and sincerity. It was natural for him to experience putting himself in his place right to the end. He took off his clothes, walked slowly into the turbulent waters, his gaze intense, and disappeared from sight. No one will ever know what he felt when he reached the bottom of the sea, but it is easy to guess what he saw. No doubt other idols, each more resplendent than the last.

Note

1. Michel Hulin, *La Mystique sauvage. Aux antipodes de l'esprit* (Paris: Presses universitaires de France, 1993).

10

A Little Theory of Thresholds in General, and of the Zero-Point in Particular

Let's imagine that God is not a given but that it is extricated from matter, like a gas. Pure, absolute, liberated, active. Karel Capek, the Czech writer, known for his play from which the word 'robot' originates (R.U.R., 1920), explored this potentiality in his novel *The Absolute at Large* (*Továrna na absolutno*, 1922) with all his characteristic finesse and grating humour. It is difficult to find in the world of fiction an experience closer to our own. In this story that is Dickian *avant la lettre*, at the very moment when atoms were discovered, a scientist invents a machine called the Karburator, which makes it possible to manufacture energy at little cost, by fragmenting atoms. To his great surprise, by consuming matter, the machine releases in the process, the Absolute, some god in a pure state.

'[L]et us assume, for the sake of argument, that God is contained in all forms of physical matter, that He is, as it were, imprisoned in it. And when you smash this matter up completely, He flies out of it as though from a box. He is suddenly set free. He is released from matter as illuminating gas is from coal. You only have to burn one single atom up completely, and immediately the whole cellar is filled with the Absolute.'[1] The Vatican implores the scientist to make sure that the production of absolute is as weak as possible, that only a kind of weak absolute is produced, minimally virulent, watered down, which would manifest itself only sporadically, because 'we are not used to dealing with the real God.

We don't know what his presence can perpetrate,' says a bishop. But without success. In the hands of a huckster, the Karburator spreads and unleashes an epidemic of religiosity. Even if 'God is not so infinite as the Catholics assert. He is about six hundred meters in diameter and even then is weak towards the edges',[2] nevertheless, no solid body is impervious to the penetration of the Absolute in its pure state, beginning with humans, who become believers and succumb en masse to mystical hysteria at all levels of society. Psychological conversion is not, however, the only mode of action of the Absolute. The release of this mad energy on a worldwide scale produces an unlimited superabundance, penetrating all manufacturing processes. It seeps into all machines, so much so that a nail factory, for example, affected by the Absolute, begins to belch out nails in astronomical quantities.

The Absolute soon takes over the 'weaving and fulling mills and in the stocking factories it laid hold on the entire textile department, and reeled off without pause millions of kilometers of everything that can be cut with scissors'.[3] It took over the metallurgical factories, the foundries, the chemical factories, the agricultural machinery and so on. 'It wove, spun, knitted, forged, cast, erected, sewed, planed, cut, dug, burned, bleached, refined, cooked, filtered, and pressed, for twenty-four to twenty-six hours a day.'[4] An extreme power, unlimited in its capacities of creation, machinic because it does not discriminate as to where it goes to stimulate production, politically illiterate because it is without any notion of political economy, the Absolute ends up triggering a global economic crisis. Incredibly flexible, capable of blending into existing religions and taking on all possible cultural faces – in Japan, for example, it becomes a very nice god, perfectly Japanese, while in the United States, it prefers gambling, alcohol and the race to break sporting records, all the while being socialist and in favour of prohibition – it soon creates multiple misunderstandings and provokes a spectacular religious war, with everyone seeking to appropriate it as their own god. The end of Capek's novel is perhaps less relevant for us. Seeking to denounce religious absolutism, Capek does not imagine anything other than an ultimate single power and does not conceive of anything other than a creator god. We are, it would appear, far from the differentiated powers of Hinduism, each with its distinct and always partial spectrum of action. Let us remember that Ganesh, for example, is not a creator god but a protector, destroyer of obstacles, promoter of knowledge and intelligence. For other benefits, one must consult others. It is not insignificant that war occurs in Capek's work at the very moment when people discover that the Absolute takes on the most contradictory faces and they begin to suffer the destructive consequences of such an indiscriminately productive power. The interest of Capek's novel for us occurs much earlier, in this strange standpoint from which the novel opens, developed further on in the mode of 'and what if ...', which consists in making the Absolute a gas or the product of a physico-chemical process. It is what remains after the dilution of matter. Immaterial, untraceable and yet, it acts.

At first glance, the challenge that Hindu polytheism takes up is not so much the invention of forms at any cost, anthropomorphic, zoomorphic or hybrid, as Hegel condemned it, ever more incarnation in the form of a body or a person, that can take abstract or disembodied forms elsewhere. It is, rather, within the very constitution, in the shaping and in the manipulation of the powers at work, to allow oneself a host of experiments that one would not necessarily permit oneself elsewhere: in a word, to allow a maximum of ontological flexibility, from the most abstract to the most concrete, from the most incarnate to the most ethereal. In other words, it is not only a question of cultivating a multitude of custom-made powers, as numerous as they are partial, starting from the principle that it is only by materializing the immaterial, by giving them form, that we will be able to conjure up everything that escapes us.

Of course, we take a fiendish pleasure in multiplying occasions for manifesting the divine, the possibilities of encounter and contractualization with these powers that are conceived as being able to keep up with chaos and to be within our reach, via a heterogeneity of supports, forms and aspects, of rites, formulas and gestures of invocation. But the opposite movement also occurs, which consists in diluting this learned technical corpus, in relativizing the rituals as being nothing but traps for the invisible, simple formulas of conjuration, and idols of the imaginary forms projected in the face of that which is, when all is said and done, in fact formless, unspeakable or inapprehensible. To affirm, in this context, that there are false divinities is, properly speaking, inconceivable. There can only be good or bad power-relays, formulas that work more or less well, techniques that do or do not prove themselves. Hegel did not see the richness of this experimental posture championed by the Hindu ritual gloss. Victim of the ontological obsession which aims at defining God independently of any interaction, Hegel sees in the plastic Hindu experiments only the symptom of a paradoxical wandering or of an imagination looking to find itself. The whole interest of *Ganesh Yourself* was to settle down from the start within this zone of fantastic plasticity denounced by Hegel and to never really leave it.

Would such an interface of divine deconstruction have produced similar effects elsewhere? Obviously not. In conditions conducive to monotheism, it would have been necessary to imagine something else, a quantum machine, for example, an abstract entity capable of absorbing all the others, or, who knows, a machine sufficiently confusing to be judged as divine, following Norbert Wiener,[5] capable of working from the blurred or the formless, from poems, novels, works of art. Nevertheless, it is in the polytheistic universes, which we have often erroneously reduced to a catalogue of ready-made and inherited forms, that we must go to look for the most elaborate means to put the effects of divinity to the test.[6] It is because God is here a floating entity, potentially everywhere, that we accept that it can lodge itself here too, expressing itself in this compound of human and machine. For the ecology of non-humans,[7] variations in the very possibilities of the incarnation of a blurred entity cannot be ignored. Before a man endeavouring to incarnate God, the reaction could have been one such as this: 'But who is this individual who takes himself for a god?' And yet, this question was never really asked as such, as though it would be more interesting to take up the game of this ambiguous machine.

One could criticize us by saying that no truly non-human conception of the absolute is conceivable in a device operated by humans who play at being God, and that the *Ganesh Yourself* experience finds its limit at this very point. Such a criticism would be to poorly perceive the metaphysical movement that has taken shape, a process of emergence very similar to the process imagined by Capek, of releasing a gas. Something comparable indeed, a chemical process, a precipitation of the absolute, takes place at the moment when the incarnant capitulates, when he or she seems to have exhausted their capacities and all the preconceived ideas of God of which a human is capable. It is then that the device, hybridizing the human and the machine to some divine composite that exists, best reveals its potentialities. Aiming at a form of 'exhaustion' of the possible, it allowed the unravelling of existing ideas of God, to imagine others which could not have emerged otherwise than unwittingly by and in and through those who took up the experiment, to produce, better than an effluent or a gas, notions that are still too material, rather an immaterial or an absence, *which nevertheless, acts.*

It remains to be identified how this release occurs, beyond the interaction between the incarnant, the interlocutor and the machine, when it happens, and why it fails when it fails. It is to precisely such an analysis that this chapter is devoted.

First, there is the discomfort in which the interlocutor finds himself when he sees a face arrive in the interface. Then there is the irony the interlocutor willingly displays when dealing with a person-object that is ultimately literal. All this suggests that God is best conceived as a formless, fluctuating power, the concretion of an unrepresentable in relation to which any countenance is necessarily excessive. This does not mean that it can only manifest itself in an abstract form, that it could not express itself through a human or that this power that we like to think of as untameable could not take control of a machine. Nothing limits either its representation or its expression, and nothing prevents the creation of new experimental conditions of appearance, since what we do by constituting it is to *propose new affordances in the world of potentials.* If a formless, pantheistic God (the God of Spinoza or Fechner) coexists here without contradiction with a multitude of forms, idols, manifestations, it is because what is present under these forms can never be really determined or circumscribed in a being.

It is a universe of potentialities which lays out for humans a boundary plane of experience, in which the invisible becomes a partner, the mechanisms at work within the cosmos become levers to be activated in order to multiply its possibilities of action. Interaction then becomes a formidable playground, an infinite reserve for a multitude of tests and experiments of generation, summoning, trapping or attraction. This is what prevents us from drawing too quickly any ontological lesson from the experiment, because the whole point of *Ganesh Yourself* could well be to go beyond the ontological obsession itself, the preoccupation with the modes of existence of beings and things in themselves, to lead us elsewhere, onto the field of potentialities, powers and effects. But to take this turn and understand the latent powers that Bappa discharges, we must try to theorize a little better the parameters of the interaction, the variations of posture from one incarnant to the other, the thresholds and the levels to be crossed in order to produce effects of divinity. In a word, to map a little more precisely the ways of generating god, or the misfires, that were encountered in the course of the narrative.

This emergent and unintentional process is not necessarily easy to grasp, to the extent that it violates the overriding intuition currently at play that says men 'create the gods', that they personify them and then animate them by ritual acts, etc. It would be better to say, that wherever one believes that gods are being made, one 'uncreates' them and that when one carries out rituals, it is so that some god is released in the manner of Capek's gas. As long as we have a vision of the problem that is more ontological than interactional, taking the two terms of the relation (or of the face-to-face interaction) as given existents rather than as things to be done, we will not only be doing the wrong anthropology of religions but we will also fail to do, with our powers just as with those of (supposedly ontological) others, concrete metaphysics. If Capek's absolute had as its main property an excessive power, favouring unlimited superabundance, the time has come to examine more precisely what exactly Bappa frees up, to circumscribe a little better than we have done up to now, its experimental power, beyond its composition, mixing up a little bit of human, a pinch of machine, an inkling of elephant.

Bappa introduces first of all an excess of incarnation or *personeity*, whereas ordinary devotion has no need. It also injects an all too human possibility of communication, where silence or tacit agreement readily reigns. This small disturbance certainly makes us aware of certain mechanisms of imputation that take place without our knowledge, but its virtue, it seems to me, is located elsewhere: it allows us to go through the range of possible oscillations and to judge, each time we freeze the cursor in a certain state, at a certain degree of plasticity and *spectrality*, the cascading reactions that this can produce on the whole invisible circuit. Take, for example, a stone, which serves as an object of worship, as little anthropomorphic as it is, and can at any moment shelter a goddess, cannot be held to be completely static by virtue of the forces which it mobilizes. Despite its apparent inertia, it is a good shifter, a passage point in a subtle network of fluctuating forces and hidden influences, ready to manifest itself at any moment. If stones, in this context, are not reducible to their materiality, it is because they carry in their very being spectral possibilities, potentialities of stimulation and convocation.[8] Living beings, animals or plants, are not the only ones concerned. Indeed, there is nothing to prevent integrating apparently inanimate objects and machines into a cosmology, objects that are vectors of a shared divine reality that can manifest itself in a plurality of different appearances. It is necessary nevertheless that machines show proof of their potential to work as *traps for the absolute*.

How to evaluate the possibilities of our Ganesh to capture (or to miss) much more than just itself, a god, a parallel reality, the mood of an entire society, even the very vibration of the cosmos? The same *kinetics of suspicion* appear to be put in place, from one case to another, in various forms. Let us summarize here the main phases that a shifter pulled off with Ganesh must inevitably go through:

- phase 1 – apprehension
- phase 2 – taming
- phase 3 – contracting.

Just like in a Turing test, people approach the robot, and try to determine who they are dealing with by asking questions. No one ever really takes the robot purely and simply as a god. It is the very opposite of naive acceptance. The interlocutor spends a lot of time testing the ability of the incarnant to properly embody the voice of God. Not everyone makes an acceptable god, of course, and that is the whole point of the experiment. The experiment does not require immediate acceptance of the fact that God can be in the machine, quite the contrary. It is necessary to *doubt* it. Very quickly, as we have seen, the incarnants formed a sort of community, with a quasi-proselyte project, ready at any moment to go to convince other people of the relevance of the device, to persuade them that there was something vital at stake in feeling this doubt, something here of a disturbing ontological experience. And the more real the disturbance, the more interesting the interaction.

The kinetics of suspicion goes through various stages, it crosses thresholds, pushed on to different degrees as a function of a given interlocutor. It often begins with the affirmation of a fundamental doubt: 'Let me warn you, I do not believe at all in what you're up to', 'I believe in God but I do not believe in the fact that he can be in there ...' or 'If you are God, why did you take this form?' and in the best of cases: 'I do not believe in all this, but since you are here, I am going to ask you a question anyway, Ganesh ...' Something of Coleridge's 'poetic faith' is exercised here, provided that it is defined other than as a 'willing suspension of disbelief', which is what it is often reduced to.[9] In many cases there is a 'consensual affirmation [or not] of scepticism', even a will to deconstruct one's own faith, but which asks only to be transformed, because it cannot remain without an answer. Once the doubt has been expressed, the challenge of the interaction is to go beyond or sublimate it.

Most of the people who questioned Ganesh were not seeking to settle on its ontological status or demystify its truth. It is the (spectral) range of its possibilities that arouses curiosity and needs to be assessed. Where God is a floating entity, potentially everywhere or nowhere, to consider that it may lodge itself here too and express itself in this compound of human and machine, has nothing to do with credulity. Rather, it is a subtle mix of curiosity and caution that invites a reflection on and evaluation of the capacity of the artefacts and devices that surround us to be carriers of divinity. Even the simplest, most modest of people, have shown curiosity towards the machine, coupled with a taste for risk, for adventure, a sense of provocation, sometimes approaching the robot with the sole intention of denouncing its mystification, but also ready to change their attitude as soon as the incarnant succeeded in arousing their interest and convincing them. Is this not the same curiosity for 'the other's gods' that historians of Greco-Roman polytheism speak of?[10]

Always on the lookout for new and previously unknown *powers*, they made curiosity their driving force rather than cultivating a stance of rejection from the outset. In the apprehension phase, initial curiosity is readily combined with caution, as there could be a trick, just around the corner. Hence the need for sometimes lengthy interactions. The incarnant must prove itself worthy of the curiosity that is expressed towards her or him. It demands from the god or from the one who claims to be its representative, signs of confidence. It is only on this condition that the kinetics of suspicion can go beyond the initial apprehension phase. And if these signals are judged sufficient, then the taming phase can begin, which also implies knowing how to exercise one's judgement. It consists in taming the novelty of what is presented to us prior to the last phase, that of contractualization, in which everything that can be done is done to try to get the recognized power on side.

We see then that one of the lessons of the experimental setup that we have imagined is to show that the 'virtues' (to use Aristotle's vocabulary) or 'affects' (to use Spinoza's) through which divine entities pass are not belief, enthusiasm, exaltation, but curiosity and prudence: as much as the first ones concern *actuality*, the second ones concern *possibility*. The possible god always precedes the actual god – and undoubtedly, the one in question is, moreover, only a modality of it, whose worth depends on the possibilities that it affords.

This is why we cannot conceive of our experiment simply as a *Turing-test-applied-to-the question-of-divinity*. It constitutes, rather, a subtle subversion of it. With Turing, ontological confusion is sought out, envisaged on the binary mode of 'human or machine', or 'man or woman', an indetermination to be resolved, an uncertainty that must be lifted and cannot remain unsolved. The experiment, when it succeeded, leaned in favour of the deceptive agent (the machine). Participants in *Ganesh Yourself* tend to agglutinate in the mode of 'and ... and ...', rather than choosing in the mode of 'or ... or ...', circumventing in a different way the difficulty of dealing with an entity that seeks to pass itself off as something other than what it is. If a relationship based on ontological confusion is never very sound, is another based on a positive conception of *ambiguity* possible?

Let's imagine that a machine sends us a message that says: 'I am a human being.' How would we react? According to attribution theory in psychology, we can take this signal at face value, accept it and behave with this machine as with a human being. But, since it is a machine, it will have difficulty hiding for long its characteristics of being a machine. It is therefore likely that we get caught up in a more complicated game and that these signals generate a misunderstanding, a conflict of 'ontological' categories (between that of the human and of the machine) or a contradiction between two 'systems of inference' as they put it in cognitive science: one that leads you to identify a human and to behave with the machine as one usually behaves with a human being, and one that tells you that it is an object, determining another type of behaviour.

We can complicate the situation further: let's imagine that you have a very strong desire to consider this machine as a human being, but that the machine periodically slips up, sending you the signal that it is only an object. You either ignore it, or you find a way or a compromise to deal with this paradoxical information, this intermittence of signals. The capacity to keep this entity holding up as a 'human being' will thus depend on the communication system in which these signals will be integrated, such that their intermittence does not lead to the stopping of the exchanges thanks to which the entity is permanently supported. Hence the importance of the precise transcriptions of the conversations between Bappa and the consultants.

Let us now imagine that a human behind a machine sends us a message that says: 'I am a god.' We can take this signal at face value, accept it and behave with this human behind his machine as with a god. But as it is a man behind a machine, we have the choice of addressing three distinct entities: identify a human and behave with it as we usually behave with a human being, to see it as a god and behave with it as with a god, or else to consider it as a machine and to behave with it as with an object. The thing in question seems simple at first sight, if one believes theories of causal attribution: one simply has to know which entity one is dealing with in order to interact with it in a fair way. In practice, however, as we have seen with Ganesh, the situation is quite different. Of course we have the choice here between three terms – a god, a machine and a human (the controller), united in a single interface (an avatar) without them ever being completely equivalent. But in practice, one does not choose, one is confronted with a *limit-presence*, which is never completely formalizable in an operator of the type 'or this ... or that ...' It obeys a spectral principle of the variation of intensities, where the entity oscillates (transcends or downshifts) amongst several possibilities. Ganesh oscillates here from the outset between the material, the organic and the aetheric.

The fact that the 'spectrum', in this case, admits three thresholds or levels (god, man, machine) perhaps complicates things a little compared to the Turing test, which admits only two (man or machine), but if it wants to be interesting for the questioner as much as for the one questioned, any interaction with Bappa must move away from the Turing-type guessing game (of the type 'you are either true or false') and put the very elasticity of the entity to the test. And this spectral range can be experienced only in the interaction, it cannot be decided in advance.

An individual wishes to be a god, a god looks for itself in an individual, a machine lets itself be possessed, allowing the two to meet. And even if the incarnant tries to take refuge behind one or the other of these entities ('I am a machine', 'I am a god', 'I am myself individual X'), the incarnant must answer the questions that are posed and manage as well as possible in this complicated ventriloquial configuration. What is primed, in this context, is what is exchanged, the capacity of the device to provide a favourable echo, answers, an intimacy of feeling. The ambiguity with respect to the being that we are dealing with – is it an avatar? How to fit this support into the panoply of existing forms of Ganesh? – is ultimately less important than the relevance of the answers transmitted by the blurred entity.

Some interactions never get beyond the initial apprehension phase; others blithely pass through it but remain interminably stuck in a phase of conflict without ever finding resolution. And many of them suddenly break down when new uncertainties arise, when the relationship is difficult to tame or when a pact is difficult to put together. It is then necessary to start all over again. These threshold effects allow us to take up again, in a different way, another experiment that has become just as mythical as Turing's: that of Masahiro Mori, known as the 'uncanny valley'.[11]

Mori's work pointed to threshold effects in man–machine interaction for the first time. This theory is sufficiently well known to be summarized very briefly: as an object increasingly has a human appearance, Mori tells us, the more that the response of the humans around it will be empathetic and emotional; but this empathy is true only up to a certain point, especially when we are dealing with animated objects that claim the status of 'creatures'. For there is a threshold at which attraction can turn into repulsion. It is this murky zone, one that is delicate to apprehend, oscillating between the 'barely human' and the 'completely human', where properties are exchanged and where the coordinates of the initial interaction can be reversed, that Mori calls the 'uncanny valley'.[12] If an entity is sufficiently non-human, Mori tells us, human traits tend to stand out and generate empathy. But if the entity is 'quasi-human', then the traits that will stand out will be those that it does not share with humans, creating a sense of strangeness in them. When the Japanese roboticist points to the tipping point at which the attribution of humanity to an overly similar humanoid becomes problematic, he imagines – as a good Buddhist – a wholly *zazen* solution to this problem that has been insufficiently explored: the possibility that another shift or disjunction will take place, of a spiritual kind, once the dilemma of the 'uncanny valley' is sublimated. Mori does not give many clues as to how to provoke such motion within (or with the complicity of) a machine. Is it a matter of tracing the properties of 'spirits', often minimal, sometimes hypnotic, in any case always well thought out? With Mori, the Buddha with half-closed eyes is never far off.[13] It is perfectly conceivable to rely on human know-how concerning rituals to conceive other types of machinic arrangements in which the invisible is a real partner, to get the *interactant* mixed up in metaphysically meaningful forms. Nevertheless, this is a second threshold, which few roboticists have ventured to explore, reducing Mori's theory to the strangeness of the initial contact, which he had pointed to, at the risk of essentializing and dramatizing it. The 'uncanny valley' has thus become, for many roboticists who read Mori a little too literally, less a metaphor for the abyss that threatens any encounter with a mutant object than a simple test that any 'effective' humanoid must pass.

Another aspect of Mori's thinking deserves our attention. Whereas it is common in modern psychology to apprehend human relations with robotic artefacts through the notions of attribution and intelligence,[14] Mori reasserts the value of *machine being* (*l'être machine*), the potentiality of forgetting itself in its act, and he questions the limit between mindfulness and mindlessness. Devoid at first sight of any reflexivity, machines afford the perfect performance of given tasks, embodying a state sought in zazen meditation, of not being polluted by thought. Mori's book *The Buddha in the Robot* is a kind of plea for a different view of the relationship between man and machine. The ambivalence of the problem cannot be emphasized enough, as attitudes towards machines always vacillate between distrust and benevolence: where some feel benevolence, others are distrustful of and fear losing their souls. How can mistrust be transformed into benevolence and vice versa? How can the latter be engaged on the basis of the former? And if man–machine interactions are above all problems of engagement (more than of attribution), we can push the reasoning further: to what extent can a machine constitute a good intercessor that digs in the invisible to release the concrete? And how then to measure its spectral effects?

This is a question that Mori left for others to explore, too busy, as he was, simulating the human rather than generating the divine. As a result, the uncanny is essentially conceived in his work as a problem of aesthetic surgery that must be sublimated or accommodated, rather than as an inevitable effect of the encounter with alterity. *Ganesh Yourself*, however, attends to this other troubled zone (the second half of Mori's schema), showing how effects of divinity can be effectively generated in an interaction. If not all technical supports are equivalent for the generation of sparks of wisdom, or divine feeling, at its high point the experiment does nevertheless make visible the possibilities of what one could call spectrosymbiosis: the machine (host entity) and the god (parasitic entity) exchange links, with the complicity of the tele-operator (the incarnant), who is also someone's shadow, the ghost of something, a hacker or pirate of God, who is in turn hackable by God or by other voices, not necessarily superhuman. But before the relationship can blossom in this mode, some stages have to be traversed.

Indeed, unlike a humanoid robot, with Ganesh one does not expect to enter into a relationship with a human, an expectation that is inevitably disappointed in the case of even the most developed humanoids. One can expect to enter into a relationship with a god, but this expectation is inevitably disappointed as soon as the incarnant opens its mouth. And most of the time one simply does not know who or what one is dealing with. Being first, the initial suspicion, tinged, even, with a strong feeling of strangeness, the interaction follows other paths, much more tortuous than the 'uncanny valley', vacillating between doubt, devotion and resignation, because once accepted the fact that the object could contain a share of divinity, it is still necessary to determine which god one is dealing with, the degree of benevolence or hostility, omniscience or, on the contrary, ignorance of this particular Ganesh. To answer, the interlocutor must measure the wisdom of the incarnant, its aptitude to generate divine speech. A real debate can then take place, and even a fight, depending on whether or not one seeks to influence the word of God: one can accept what is said, but disagreement can also be shown, to try to convince the god that one is right or else to seek to influence it, to nourish it with new principles. With such a god, it is not surprising that there was much talk of politics. This was the sign that engagement had been successful.

Each interaction draws its own curve, a kinetics of doubt that is singular to it (phases, clutches and thresholds). God allows himself to be more or less subdued or tamed and sometimes the divine presence emerges all the more strongly when it does not allow itself to be dictated to by the terms of the contract and imposes its own. The balance of power is punctuated by thresholds (more or less crossed) and effects (more or less mastered). If we want to go beyond the model of the 'uncanny valley', or at least enrich it, it is also necessary to manage to map the regions

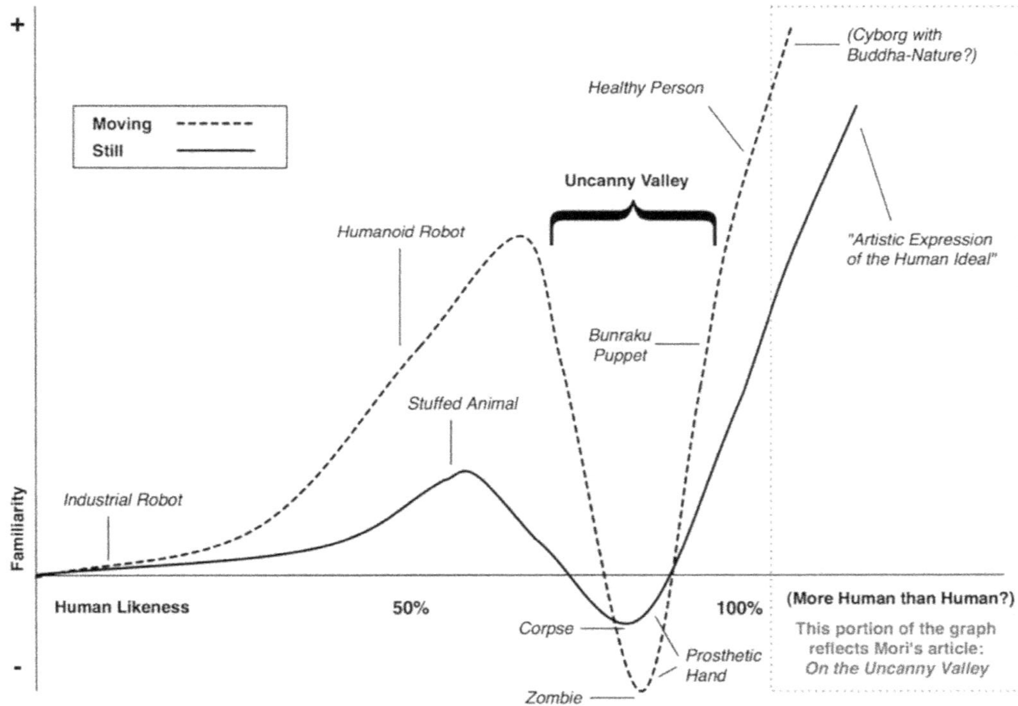

Figure 10.1. Mori Masahiro, 'Bukimi no tani gensho 不┌味の谷現象' ['The uncanny valley'], *Energy*, vol. 7, no. 4 (1970): 33–35.

Figure 10.2. An attempt to visualise the 'uncanny valley' problem differently.

which are situated in the prolongation of the valley, explored with more or less depth according to the singularity of each incarnant, and that of their interlocutors, and which were not identified or located by Mori, because these zones of turbulence are outside his schema.

The Theory of Thresholds

First of all, there is a *threshold of strangeness* (STRANG), which is equivalent in our case to the discomfort caused by the confrontation with an idol with an all-too-human face, functioning in an unusual mode, despite any impression of recognition or an initial affirmation of divinity. This threshold is usually crossed, but it sometimes happens that the interlocutor remains stuck on a plateau of discomfort for a long time or that the discomfort is never really overcome.

The *zone of ontological ambiguity* (ONTO) designates this zone of discomfort and doubt. Overcoming the initial doubt is most often the condition for continued dialogue, but doubt can return cyclically and both 'ontological confusion' (OC) and its denial are also drivers here.

The threshold of the meta (META) refers to the way in which the interface gains in *metaphysicality*, transforming its immediate exceptional appearance into a lasting divine manifestation or else, if a 'cosmicity effect' (CE) proves unattainable, in order to persuade the interlocutor that it is more than just a human trapped in a machine. This is a 'clutch' of the divine, usually possible only once the 'threshold of strangeness' is surpassed and the 'zone of ontological ambiguity' traversed.

The *darshanic area of attraction* (ATTRACT) derives from a principle of interaction (called *darshan*, the Sanskrit word for view, or sight) at work with stone idols, yet concerns not only those beings worthy of being seen. By virtue of this principle, a divine manifestation, whatever its form, has in itself a beneficial efficacy simply by its very presence. During the Ganesh festival, for example, a good devotee will endeavour to see about twenty deities in a day. That there is a benefit to be gained from its mere presence (with the granting of a wish at the end) can be claimed by the incarnant and refused by the interlocutor when the latter denounces it as an impostor, or vice versa when the interlocutor

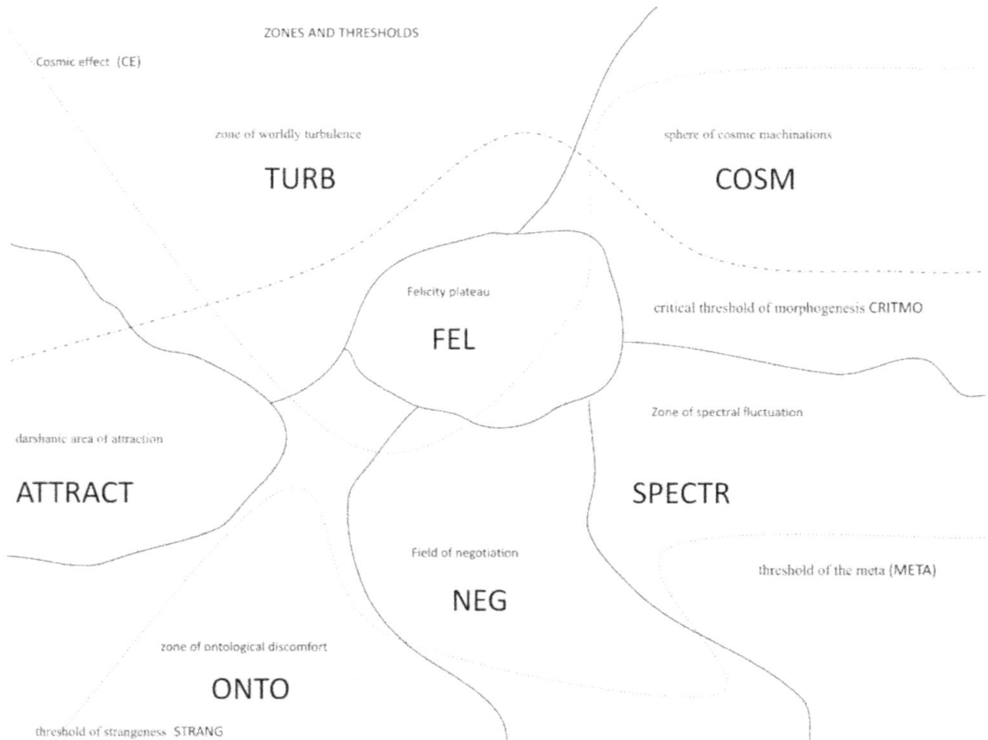

Figure 10.3. Theory of thresholds. The 'uncanny valley' replaced by a more chaotic landscape made of multiple (and fluctuating) zones.

asks the incarnant for its blessing or else when they see it as a divinatory interface. Whatever happens, this comfort zone often has a hard time settling in, and if it does settle in, it must be overcome if a constructive dialogue is to be achieved.

The *field of productive negotiations* (NEG) designates another zone about which it is up to the interlocutor to evaluate the extent, and which depends on the propensity of the god to let itself be used, its degree of manipulability and the interlocutor's need to make demands, both on and before a god who can be the vector of ideologically connoted ideas. The opportunity offered to the interlocutor to bargain over a multitude of things, to take up a number of subjects or to enter into a power struggle is all the more manifest, since the interface comprises a higher degree of interactivity than an ordinary idol. One can seize the opportunity to try to obtain something, have a wish granted or ask a question one cares about, but one can also put in question all relations with the deity in this form, inviting a reformatting of the ordinary devotional relationship, or else dictating the best way of being a god.

The *zone of worldly turbulence* (TURB) is a zone of intense conflict but also of the redistribution of the power to act. There is, it seems, a threshold of critical turbulence which varies from one incarnant to another and which depends on his or her capacity to solve the questions which are posed or the impossibility of getting himself or herself out of this zone. Crossing this threshold is essential for power to be productively redistributed and for new pacts of devotion to emerge.

The *sphere of cosmic machinations* (COSM) refers to the alliances the incarnant is likely to form with other powers (from ecology to the nation, from the stars to karma) and how the interface will synchronize with something other than itself to constitute a cosmic machine. While the cosmicity effect is rarely the result of immediate perception, the extent of possible cosmic connections is the ground on which the god is ultimately evaluated. Making the right alliances and demonstrating in real time a deep synchronicity with the cosmos are crucial for peak cosmicity to be reached.

The *threshold of critical morphogenesis* (CRITMO) designates the fact that even if the interface is related to the deity, it is up to the incarnant to demonstrate filiation, to fit into a zoomorphic lineage, to find its function or to carve out its interstice in the pantheon of existing gods. Singular gods may emerge, once this threshold has been convincingly crossed, lines of speciation totally out of step with known gods, or those that are unidentified, often unbeknownst to the incarnant himself, who may be surprised to initiate a new lineage.

The *spectral fluctuation zone* (SPECTR) designates the possibility for Ganesh and the incarnant to meet in a completely different mode, to repolarize themselves, despite disagreement with the incarnant or unacceptance of his divine status, in a new geography of incarnations, between the material and the invisible. The god is in fact all the stronger when he is expelled outside the machine, as when the incarnant is judged negatively, for example, but Ganesh can also multipolarize himself in another way when his electronic manifestation is judged ambiguous but potentially acceptable.

The *felicity plateau* (FEL) designates any form of modified state of consciousness, of wisdom emerging from the interaction. The dialogue can indeed change the way the protagonists see the world, whether it is about a metaphysical idea, the nature of God, the right way to practise one's religiosity, or practical solutions to political or ecological problems. A *clairvoyance threshold* (CLAIR) is reached when a conversion takes place, when a new possible contract lets itself be shown, satisfying the protagonists and making it possible to overcome the violence of the initial relationship, to attenuate the vivacity of a demand or to overcome the state of frustration produced by the impotence of the deity or else a possible disagreement. But sometimes it also happens that no agreement is found as to the right redistribution of powers that should be implemented, no real conversion. And in this case, we fall back into SPECTR, where the god exists all the more because not incarnatable.

Let us try to see all these zones of turbulence and thresholds at work in practice. It is necessary to have a view of the situation that is more chaotic than linear, because they do not necessarily follow one another. It is a question of better grasping the *fields of disturbances* through which the god must pass in order to appear, mutate or disappear before resurfacing in a new form, but these fields can be articulated in various ways, which we will try to show. It is quite possible, for example, for the god and his interlocutor to go back to a previous zone after having crossed a threshold, or even to start again from zero, at least in the most conflictual interactions where the interaction itself acts as a power in and of itself, an infernal cosmic washer. Moreover, each zone has its own landscape, when one examines the course of the interaction in detail, with its fluctuations, peaks, plateaus and drops.

1. STRANG/UNCANN → ONTO → META

Let's start again from the first threshold. From first contact, the interface produces an effect of strangeness (STRANG). It gives the impression of being quite similar to an ordinary idol, but it has an animated and all-too-human face. And as soon as the incarnant is noticed and begins to speak, the expectation of divinity is thwarted and an even greater zone of ontological discomfort (ONTO) is entered. Once initial doubts have been expressed by the interlocutor, the incarnant must conquer its deity by whatever means, which will often take a long time. The discomfort of not being understood, of being misperceived or of not being divine enough is there to be overcome by the incarnant. And it is usually overcome in several ways. Karan, the first incarnant, faced this problem from the start, trying to reinforce his divinity with considerable energy, invoking various possibilities: being an electronic Ganesh, or incarnation straight from Mount Kailash and so on, without ever quite winning over his interlocutor, and in fact ending up facing exponential suspicion. Some cling to these initial scenarios as if they were lifelines, but soon realize their limits.

A whole stock of shared mythological episodes were mobilized, from the start, to produce a desired effect. But other characters, such as Anand, Varsha or Jyot, were driven by their interlocutors to invent shifted, novel tasks, loosely linked to already-known narratives. This is proof that the threshold of the meta (META) cannot be reached easily. Searching for it usually triggers considerable effort on the side of the incarnant, but experience shows that it is best crossed accidentally, involuntarily, at the moment when the incarnant, exhausted by questions from its interlocutor, from its efforts to make itself acceptable or from its failure to win over its interlocutor, gives up any effort at attempting incarnation.

It is necessary to make a distinction here between divinity proclaimed from the start, and divinity that is slowly gained or that emerges without the knowledge of the protagonists. The initial

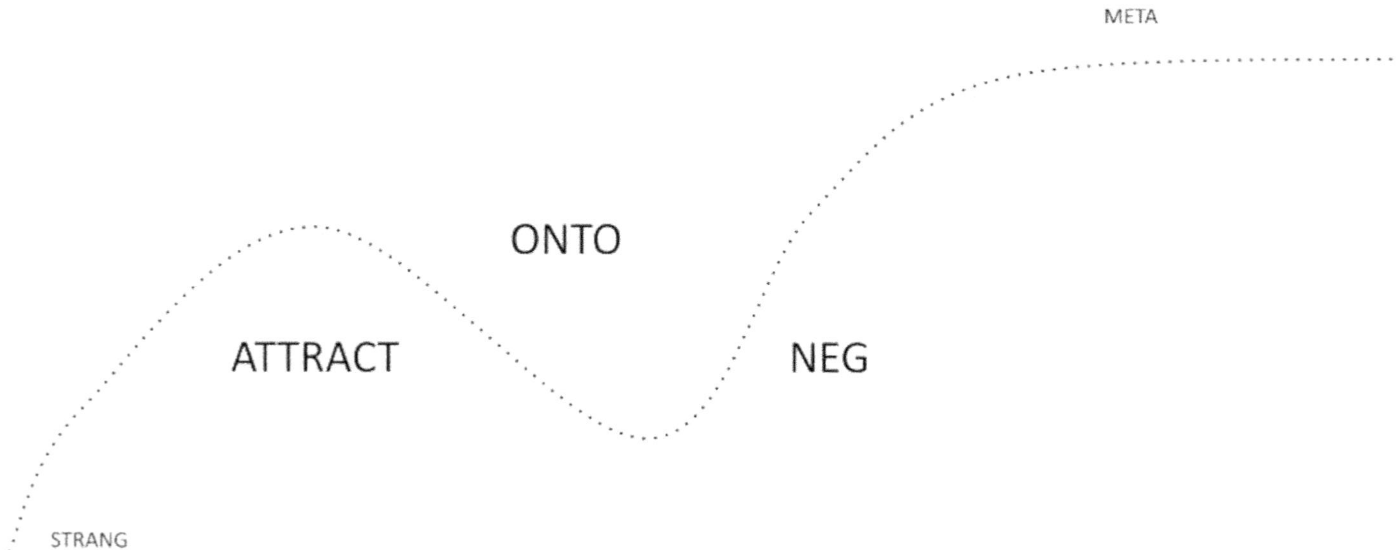

Figure 10.4. From the strangeness threshold (STRANG), which designates the discomfort caused by encountering an idol with an overly human face, to the meta threshold (META), which refers to the way in which the interface ascends to metaphysicality, and the incarnate conquers its status as a worthy contender for the divine.

affirmation of divinity generally increases suspicion more than it attenuates it. It is very often perceived as an abuse of power. Karan tries to take advantage of the resemblance he might have to an idol, at first blush, for his interlocutor. He is, however, quickly brought back by his interlocutor to his fundamental ambiguity (ONTO). She asks him to explain his novelty, which gives a troubling impression of a threshold of strangeness (STRANG) that has been badly crossed, of a threshold of the meta (META) that is impossible to reach and of a zone of ontological ambiguity (ONTO) from which one cannot get out, which drags on or which is just one vast plateau.

2. ONTO ← ATTRACT → NEG

That the ordinary idolatrous relationship (ATTRACT) struggles to get settled, or to last, despite Karan's attempts at hypnosis, and that the interlocutor always brings Karan back to his ambiguity (ONTO), together illustrate the chaotic movements, and the pressures that are exerted, introducing hesitation into the interaction. A shift of zone into another region of our diagram occurs, but without engagement in any straightforward way. This does not mean that a *darshanic* wind of attraction will not blow again, if I can put it this way, colouring the interaction like a cloud (or a gas) of the absolute. The incarnant seeks to reinstall it in one way or another, seeking to take advantage of it at the right moment. Darshanic attraction, however, meets strong resistance here.

It is the same with Nitin, the bus driver. He does everything he can to resist Karan's attempts at hypnosis. Nitin does not wallow in any zone of ontological discomfort (ONTO). He accepts the ambiguity of the entity from the outset and wants a confrontation to the point of war, no matter who he is dealing with. The field of negotiation (NEG) is established from the outset. It presupposes the acceptance or even temporary suspension of doubt, of suspicion. Nitin refuses to allow the dialogue to be polluted by a false debate about the status of the interface, because he knows that he is dealing with anything but a god. In order for such a transactional ground to be established, and for the extent of the bargaining options to be assessed, there must be a strong desire to obtain something, to formulate a wish or a simple question that prompts negotiation with the deity. Thus, in this case it is a productive shift from the initial zone of discomfort (ONTO) to another zone of turbulence dictated by a balance of power. In Nitin's case, however, where there was no initial discomfort and in which there was a desire to engage in unreserved debate, another tectonic shift takes place. It is from the outset that Nitin launches the incarnant into the zone of mundane turbulence (TURB) and it is within it that he settles down to try to gauge the extent of any possible negotiations (NEG).

3. ATTRACT ← NEG → TURB

The effort required in order to get out of an instrumental or oracular relationship to the divinity is the object of endless negotiations, bifurcations and aborted attempts. Don't think that the initiative to get out of such a rapport always comes from the interlocutor. Anand, for example, refuses to be taken for an oracle or an interface from which one could obtain something by magical means. That being said, in the case of Karan's first dialogue, it is his interlocutor who pushes him to a radical questioning, denouncing the vicious circle of devotion, and it is Karan who finds himself having to negotiate his own existence. He has no other way, then, than to multiply promises, trapped by his own superfluousness. It is the same with the bus driver, who denounces the ineffectiveness of mantras or meditation as an escape from real problems. To refuse the *darshanic area of attraction* (ATTRACT) is to renounce a whole economy of devotion. The mechanisms that then pull us away from the mundane devotional relationship, and away from ordinary transactional possibilities, are most often the consequence of the interlocutors refusing the easy solutions offered to them.

Bringing God up to date, adapting it to modern times, is much more a request from the interlocutors than an initiative of the incarnators. Indeed, in most cases it is the interlocutors who ask to leave behind the relation of devotion-reward, meditation, the ritualism of the mantras and the oracular relationship, to invent another pact, more in line with the reality of the worldly turbulences they are going through. And when the incarnant takes the initiative on his own (as Anand did), he becomes the relay point of a demand to recast this pact, a demand widely shared by other interlocutors, at other times. Faced with the refusal of proposals for a classical style of devotional contract, the incarnant is forced to look elsewhere for the bases of a new posture to be taken. And because the incarnant does not find such a foundation, he gets lost in repetitions, stammering, or else has no choice, like Karan, but to beg his interlocutor to grant him a period of reflection, to temporize, in order to regain the means to carry on the negotiation.

4. TURB → COSM → CE

Under the full force of the turbulence imposed by his interlocutors (TURB), Karan hesitates to enter the field of negotiations and takes refuge behind the machine. We see Anand doing the same thing when he invokes his status as a human prisoner of a machine. Each one endeavours to invent a legitimate posture for themselves, seeing themselves as either the victim or the agent of a machination. We then enter another sphere straight away, which aims to constitute the most exuberant cosmic machines (COSM). It is possible that, in the case of Karan, the impression that we are dealing with a strange pact between the gods and the machines is amplified all the more as he chants the same formulas over and over again, in a mechanical way, sometimes even pretending to ignore what his interlocutor is saying to him. The *machine language* that Karan finds, composed of opinions and theological snippets, synonymous with the cooling off, or drying up, of the divine word, in the end opening onto a lack of empathy, is none other than a way of giving form to a machinery of greater magnitude. A god-program, both insensitive and in retreat, of which he would be only the last avatar.

To occupy the COSM, Karan is forced to adopt another strategy. He begins to pronounce the syllable 'Om', which takes on a completely different tone than when it is pronounced in a ritual context. There is synchronization with something other than himself, a will to engage the interlocutor in a parallel universe, another dimension, a matrix, an invisible network, another cosmos. The mantra unites the one who formulates it with a physical, cosmic, mineral reality that precedes the emergence of living beings, a milieu where vibratory forces are spread by resonance and propagation. For every Hindu, the sound, the

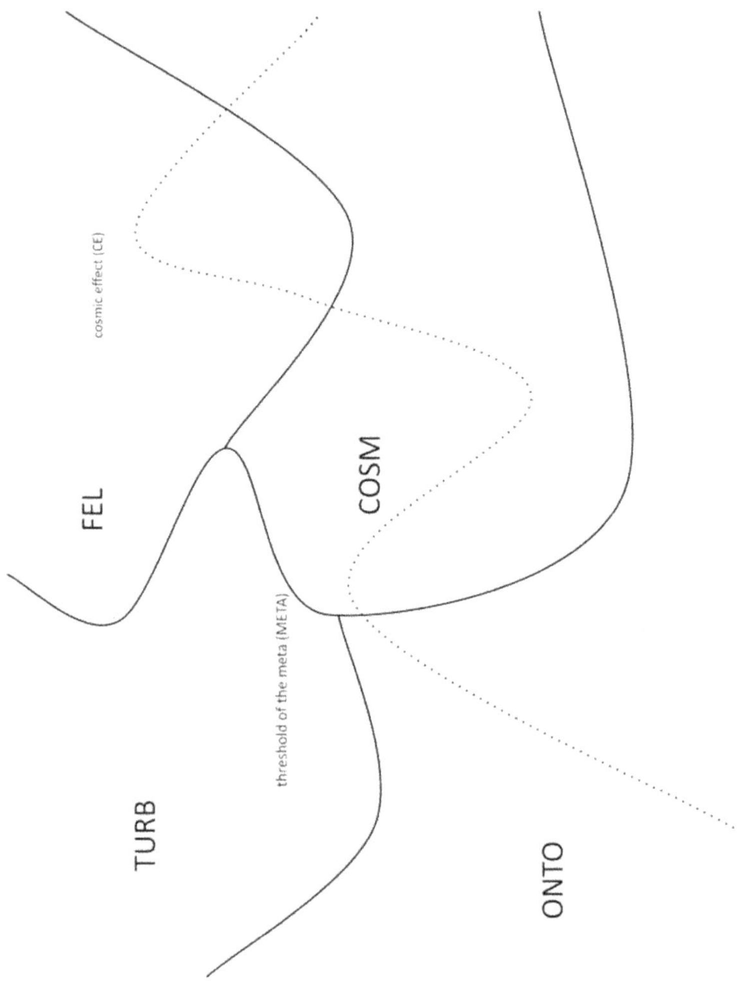

Figure 10.5. From ontological discomfort to a cosmic effect, the incarnator and his interlocutor go through zones of turbulences.

voice, at a certain frequency, has this power of connection to something greater than oneself. But this time, Karan's interlocutor can see only a desperate attempt to flee from the magnitude of the problems he is asking him to solve. A good god cannot miss the synchronization with the world of his interlocutor (TURB). He must first synchronize with things that are tangible and whose effects can be measured. This is the message Nitin is trying to convey.

It took Anand multiple failures, running into the irony and scepticism of his interlocutors, to realize that displaying oneself as a human like any other was in fact less risky in order to have a hope of patiently crossing the threshold of the meta (META), than to claim an imaginary cosmic circuit (COSM). Based on such humanization, the incarnant could expect to generate the trust that had been missing until then. Anand had always come up against the fact that there was a residue of strangeness in the interface, which the interlocutor could not make sense of, preventing him from fully situating this new phenomenal form in the panoply of known forms. Everything happens as if engagement in a more productive zone of debate could not occur as long as the incarnant did not find his form or a morphogenetic lineage acceptable to his interlocutor, a point that will be explored further. If it's a human, why then is it in a machine? If it is a god, why then take the form of machine and not simply a human form? Probably because of his experience as an astrologer, it was Modi who understood that the best way to fully inhabit the COSM was not only to make an alliance with other guiding powers, such as karma or the stars, but also to make a pact with uncertainty and the play of probabilities.

In any case, synchronization with other forces is never acquired once and for all and has to be done several times. Sometimes the incarnant claims to be in close proximity to a greater course of action, whether destructive or creative (Anand), the great movement of national liberation (Talpade), or the ecological struggle or the struggle of particular groups. Bappa thus very quickly became the ally of minorities, gay or feminist activists. Some proclaimed alliances triggered strong protests. Karan clearly sides with the state and the powerful when he is asked to defend the oppressed, abused children and those who want to take justice into their own hands against the corrupt. Other alliances must be forged, and some are woven in real time, simply by seeking to connect: after feminizing Ganesh, Jyot asserts the privileged occult relationship she has knotted with an invisible matrix, through random appearances on the internet ('Hello Hello Ganesha?'). All those who have used the machine to send messages all over the city are each in their own way exploring the possibilities of a matrix alliance with a larger circuit in order to constitute a cosmic machine, suggesting that after these first experiments, a multitude of Ganeshas could well appear, by the most diverse technological means.

5. COSM → SPECTR → CRITMO

Once the interaction has started, there is no way of stopping its flow. The incarnant, no matter what it does, is allied to the divinity, even if it is a fuzzy alliance, which is questioned, rejected as unacceptable, or which necessitated being redefined in a new spectral geography (SPECTR). It happens very often that the incarnant feels temporarily trapped or stuck, but even the imprisonment is experienced, perceived and interpreted as a state in a process of incarnation, the end of which is not known in advance. When he realizes that simply proclaiming himself God does not work, the incarnant is forced to resort to other tactics, to invoke other forms of kinship with divinity that make his manifestation as a machine plausible or acceptable.

It should not be assumed that alliance within the spectral fluctuation zone (SPECTR) is necessarily more stable when the status of the incarnant is clearly defined. In the end, the alliance becomes more stable when the incarnant has difficulty in gaining acceptance of its form, and thus fluctuates between several possibilities: an alliance that provokes criticism or mockery inevitably triggers the search for other pairings not considered by the protagonists. Each 'exit from a role' is an entry into a 'future role'. Once the game of incarnation is accepted, it is never really possible to extract oneself from the Ganesh matrix, or pantheon. The incarnant is forced in spite of himself to find his place in it or to prolong it in one way or another, to find his singularity, to reach, at the risk of failure, a critical threshold of morphogenesis (CRITMO). One might think that it would be in the interest of the experimental set-up that people not make a difference between

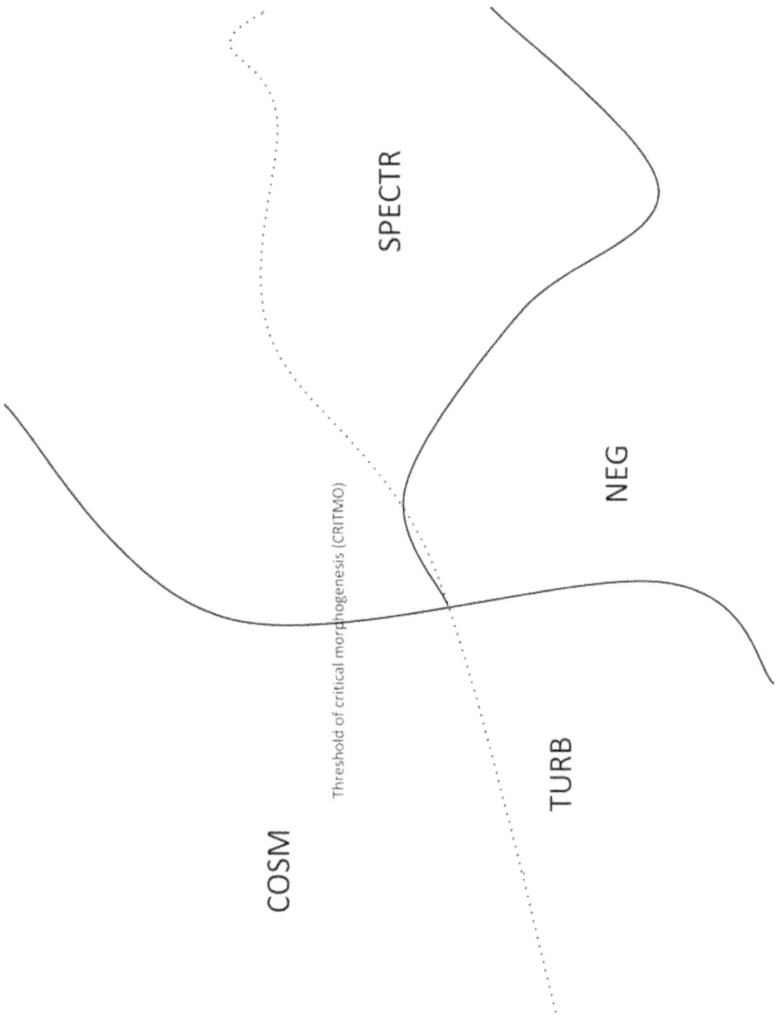

Figure 10.6. The path to morphogenesis.

the entity and its manifestation, that the interface make people forget questions about the nature of the beings that are there. This ontological confusion is, however, rare. It is, rather, doubt about the nature of these beings that immediately prevails, preventing what one could call a zone of perfect illusion (ZPI) from being established. Sometimes, though, the interlocutor needs absolutely no clarification to play the game. This was the case with Sushant, for example, who admits that he was seduced by the interface as one might be by a beautiful idol in a temple. The spectral fluctuations are all the stronger when the interlocutor wonders 'who he is dealing with' or 'who is there' and the exchange is hampered by the confusion of the incarnant's answers. But most of the time, no one is fooled and can only let himself be drawn with more or less reluctance into this uncertain space where each person has to evaluate for themselves the nature of the manifestation he or she is dealing with, and the possibilities afforded by the simulacrum.

The way in which the constitutive absence (of Ganesh) is negotiated throughout the interaction can follow several paths. One might think that the more successful the incarnant is, the more Ganesh is felt, the clearer the location of the god is, but in reality the failures of incarnation contribute to reinforcing his spectral regime of existence, his 'non-incarnable' dimension. Let us emphasize that the failures from this point of view are much more numerous than the successes, but the interest of the experiment lies in the fact that even a failure of incarnation always ends up being overcome and ends up reinforcing the regime of existence of the entity. It is its flexibility that is the pivotal element of the experiment and the experience. It should be noted here that *Ganesh Yourself* leads to a very different model to Mori's 'uncanny valley', where the pivotal element of the interaction is the degree of confusion between the human and the machine.

In *Ganesh Yourself*, what matter are the 'spectral' possibilities of the entity, the flexibility of the association and dissociation of its components (machine, human, god). Indeed, the god can persist further in the spectral state as soon as it is considered to be badly incarnated. A repolarization can occur which vacillates between three great possibilities: either God is entirely diluted and 'there is no god here', as Anand said, he is only a pure play of mirror effects, reflection or reverberation of something which is in the head of the interlocutor, or God is well and truly there in the machine in the way that an entity can possess a body, or finally, God is expelled elsewhere.

The expulsion can take several forms: either God is rematerialized in the form of a 'mental image' (Karan), relocated in the inmost depths of the psyche, or it is expelled far outside the interface, it is 'non-incarnatable'. When it is the latter scenario that takes over, the interlocutor has the impression of being faced with a disembodied face, a shadow of Ganesh that risks fading away at any time, to disappear or to be extinguished (Anand, Karan). The incarnation loses resolution in relation to the invisible, who is behind it is inaccessible to perception. But, more often than not, by dint of arguing or because he struggles to form a firm idea of what he is dealing with, the interlocutor does not bother to fix the confused entity at all costs. Its undecidable status is not an obstacle to dialogue and there is a composition of scenarios implying a strong degree of vacillation of the entity between several 'localities' and regimes of materiality.

6. SPECTR ← CRITMO → TURB

In the serial prototyping of Ganeshas that constitutes this experiment, one of the major challenges encountered by the incarnant consists in individuating himself as a god different from the others, in finding his morphogenetic line (CRITMO). He is generally pushed by his interlocutors to specify himself more and more, whether it is at the level of his form (in the name of whom and of what he is the incarnation) or of the content of his speech, which must differ in one way or another. He can assert his uniqueness, the rarity of his form as well as the originality of his opinions, or, on the contrary, he can melt into the characteristics of an already existing image. But it is often precisely to the degree that an incarnant is forced to say something singular or not in conformity with the divine word that the exchange is worth pursuing. It happens in certain cases that the incarnant creates, together with the machine, a kind of enigmatic and ambiguous assemblage, whose worth stems from its irreducible singularity. This was the case with Modi, the fiendish astrologer, with Jyot, who took advantage of the exercise to demasculinize Ganesh, or with Varsha, who made of it a god who was not a destroyer of obstacles but, rather, a creator of them.

In all cases, when the incarnant finds their own threshold of critical morphogenesis (CRITMO), that is to say, the novelty that gives meaning to his incarnation, he resolves a formidable tension that has been established between two contradictory tectonic thrusts, one which consists in fleeing into the ethereal spaces of the cosmos – escapism is a tendency which crosses the two regions of the COSM and the SPECTR – and the other which consists in being as grounded as possible, taking worldly turbulences (TURB) in hand. 'What is heaviest, you heroes? asks the spirit that bears burdens – let me take it upon me and rejoice in my strength,' said Zarathustra's camel, ready to carry all burdens. There is no effect of divinity possible without an increase in the load to be borne. Modi says it in his own way: a god must have an answer for everything, otherwise he cannot claim this status.

In any case, the ability of the incarnant to respond is bound to vacillate in the face of the magnitude of the problems that are submitted to him. Whoever refuses, for example, to express himself on topical issues attenuates his burden, but he is readily accused of irresponsibility or indifference by limiting his area of intervention. Not to be indifferent implies accepting to take the most banal as well as the most violent questions that are addressed to us, and to look for a solution without necessarily sending back to the interlocutor a message concerning the impertinence or the absurdity of his question. The interlocutor is not the only one to engage in real trials against the impotent god.

The incarnants, on their side, whatever their political ideas, have sensed that the interface must be cathected. If God must manifest himself in the appearance of a machine and if there is some legitimacy to this 'technological' form, it is because a revolution must take place that practises change. Bappa quickly became a prophetic interface, supporting multiple causes, from ecology to the fight against radiation (Kumar). The incarnations followed one another to announce the worst calamities but also to propose solutions. 'To each Ganesha his cause, to each cause his Ganesha,' said Anand. The success of the interface was not only due to the fact that it made visible in a much more explicit way than usual the relationship to the deity as an open game of delegation, a transactional exchange, involving actions on one side and results on the other. It made it possible to envisage a redistribution of the burdens that weigh on each other, new effects of discharge and of taking charge in the mode of 'You god, what can you really take charge of?' and 'You human, of what are you ready to unburden me?'

7. TURB ← FEL → CLAIR

Let us note that in the turbulence of negotiations, the incarnant's objective is always in one way or another to convert his interlocutor. This objective, however, goes both ways, to convert the god to an idea or to dictate to him the right way to be a god when he realizes that he has not internalized his role well. Nitin Pawar, the bus driver, for example, tries to instil in the god a sense of reality and to inform him of the situation regarding education and corruption. The dialogue is then punctuated on both sides by small conversions, sometimes successful and sometimes unsuccessful, and by bursts of resistance, because it is rare that both the incarnant and the interlocutor accept the proposals made to them without endless debate, whether it is a proposal concerning the phenomenal form, the right way of practising religion or the need to reform it. To let oneself be dictated to about the right way to be a god supposes being willing to accept a certain amount of ventriloquism, to let oneself be manipulated by the interlocutor. 'I am what you make of me,' says the god after desperate attempts to reach an agreement. 'I dictate [a message] to you and you transmit [it],' says the interlocutor, or, 'you listen to me and I obey you', are reduced, synthetic statements of the variety of tensions that punctuate encounters, when it is a question of making the interface amplify or diffuse messages. What good is a god if he is not a good amplifier, Nitin suggests. But the incarnant can also resist the ventriloquial desire of the interlocutor (Karan), can assert his intractability before losing it and then regaining it.

How to evaluate the propensity of the device to unwittingly generate a clairvoyance effect (CLAIR)? Wisdom can be attributed by the interlocutor to the incarnant, but the incarnant can also have its awareness greatly expanded through contact with an interlocutor, when he comes across one wiser than himself. Clairvoyance can take several forms. The affirmation of a metaphysical principle, a good word, a positive thought, a cheap humanism, can either carry along belief, or else arouse admiration. Clairvoyant speech, provided of course that it is judged as such, suspends any questioning of the status of the incarnant. Something divine is transmitted through the interface and it is its capacity to transmit that is judged. But there can also be disenchantment with anything that might be of the order of positive thinking or even of 'belief'.

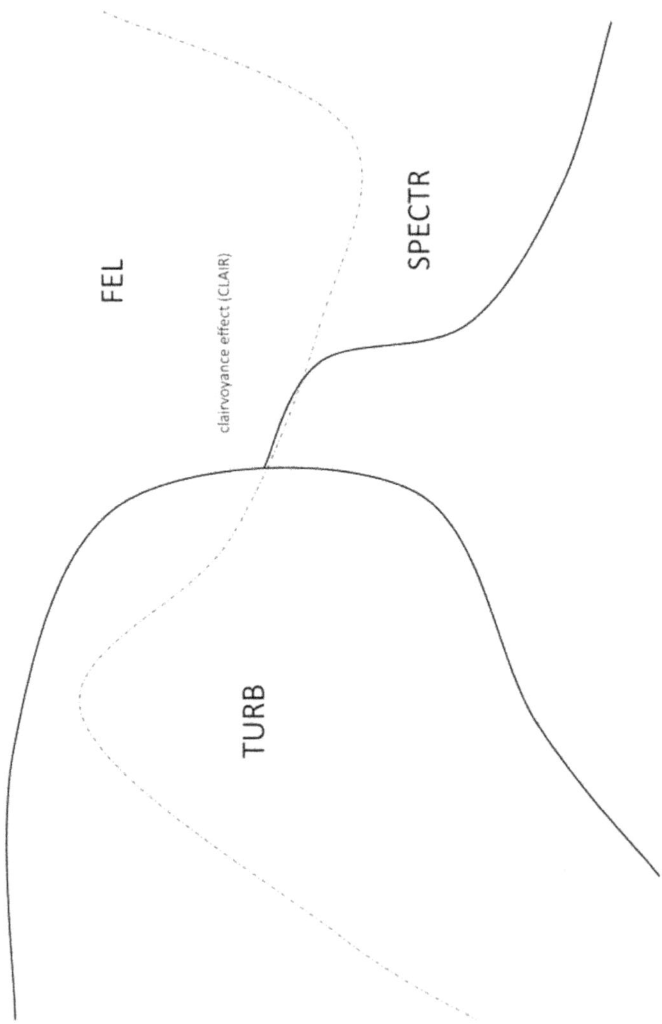

Figure 10.7. A way to clairvoyance.

A condemnation of devotion practised in a magical way can be experienced as an 'expansion of consciousness', or the disruption of a preconceived idea. Well-known Indian gurus (of the Osho type) have made extensive use of this kind of demystification and disenchantment in order to disrupt their audience. A prophecy, unverifiable or not (as in the case of Modi), can have a stimulating effect on the emergence of clairvoyance. Modi uses the most dubious prophecies to generate flashes of awareness in his interlocutor. Professor Kumar also announces the most terrible catastrophes, even if they are more scientifically based. In the same way that the incarnant must come out of himself, galvanized by his own prophecy, Professor Kumar's call to action is a cry thrown into the public sphere that takes on a prophetic form in order to provoke a change in the interlocutor's behaviour, a radical questioning of the routine.

To conclude this overview of what it is that *Ganesh Yourself* allows us to explore, beyond the 'uncanny valley', let us note that the effect of the device is always to operate a requalification of powers, even down to the matter of their redistribution. The *felicity plateau* is truly reached only when the directive power referred to under the concept of God is effectively re-evaluated, re-actualized according to other forces acting in the sphere of worldly turbulence (radiation, climate, corruption etc.) and is then distributed differently. When the interlocutor flatters the god with all the necessary marks of respect, the incarnant may feel a sense of omnipotence, but it is rarely tenable or lasting. Most of the time, it is the very form of the god-person who is called into question, or even sometimes the delegation to an invisible omnipotence which is savagely devalued as unlivable by the interlocutor, who is in search of another relation to the universe of powers. One should not believe that it is when the directive power is redirected towards the human that the effects of emancipation are the strongest, when there is, for example, an invocation by the incarnant of the 'superhumanity' in man or of the 'divinity in oneself' (Anand and many others endeavoured to defend this kind of statement). Conversely, the directive force can be expelled into the depths of the cosmos, under the pretext that men have no power, and suddenly abandon the divinity and the one who claims it, denouncing both as powerless.

The redistribution that takes place is often complex and power is rethought only after a creative operation of fragmentation or decomposition. Karan, for example, in his interaction with Nitin, considers that it is only the power of creation that belongs to God and that the power to act is in the hands of men. Defending the idea that only men have the key to solving their worldly problems, he 'disempowers' himself, displaying a Brahmanical detachment. But Nitin does not want to break with the idea that God is political, that God is an integral part of the world, and a virulent debate ensues about the powers that really act in the field of worldly turbulence. Despite the best efforts of both, the interaction ends with an inability to find a satisfactory compromise. But even in this case, where the protagonists struggle to find an agreement, a state of felicity (FEL) can occur, due to exhaustion, because they have reached the limit of their capacities to creatively rethink God. It is even to be expected that a plateau of felicity will be reached in two seemingly contradictory ways, after a radical critique of all personifications, the questioning of the very mechanism of divine manifestation (as between Karan and his first interlocutor) and the observation of all impotencies, or, conversely, by new personifications as when Nitin condemns greed as a new god, when Modi makes a timeline a divine force, the one that he himself embodies, or when activists use the interface to generate super-personifications allowing them to act within a requalified and re-actualized field of forces (radiations, climate etc.)

Zero-point

The model proposed here helps to grasp how far, to which abyssal depths, the interaction can go, but it certainly has its limitations. I see at least three of them:

First, it is more complicated than the 'uncanny valley', since it amounts to multiplying the regions that are extensions of it. A two-parameter curve that varies, for example, *inertia* and *the living*, or the *faculty of speech* and *divinity*, would certainly have made it possible to point out a paradoxical tipping point in the way that Mori does. The more we are inert a priori, the less we are alive, but there is a point where this is reversed, when we reach a certain threshold of dematerialization, an inert entity can endow itself with a movement that is more alive than in the living. The more convincing our speech is at first sight, the more

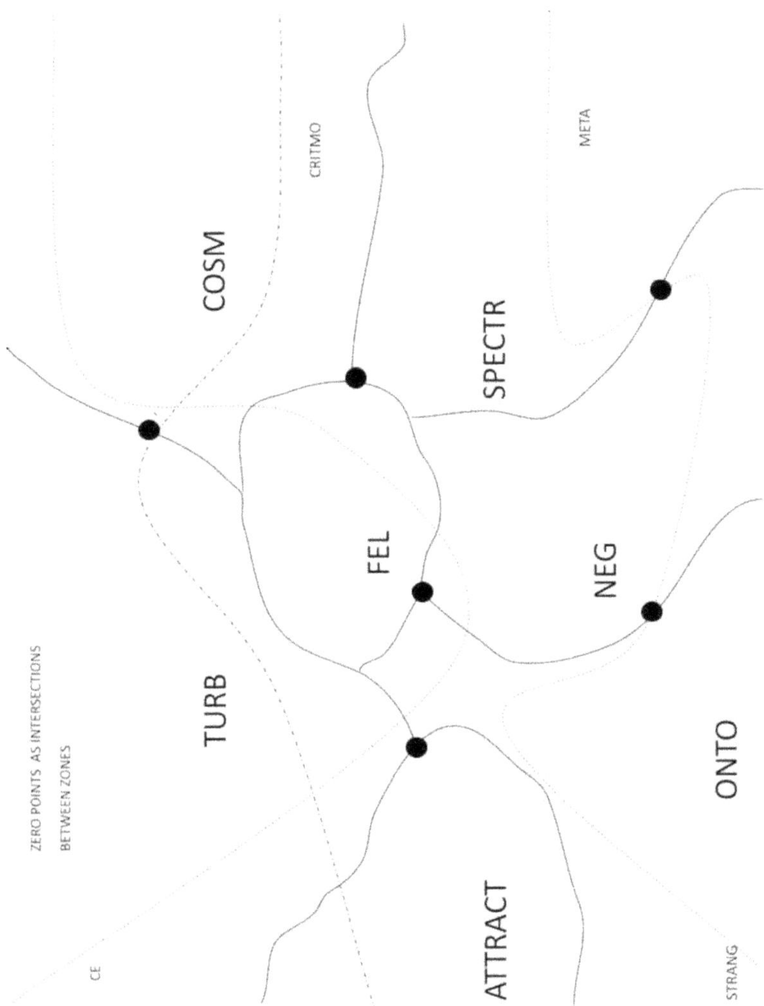

Figure 10.8. Zero-points as intersections between zones.

we can prove its divine power and this can be reversed by dint of debates, after having reached a point of saturation, the god being expelled into the invisible. These movements, it seems to me, do not produce a zone in their own right, they are, rather, internal to the different regions, and in particular to the region SPECTR for the first (concerning materialization and liveliness) and TURB for the second (concerning the divine power of speech).

The regions or the *zones*, then, because they are the result of intense and turbulent exchanges, are not watertight, they do not follow one another in a linear way but are intensely chaotic. As I said, the interaction puts both the incarnant and God through a kind of washing machine motion, pushing them both to a point of singularity or catastrophe where they are forced to reinvent themselves in their relationship.

The third limit is that thresholds are all the more delicate to identify as they have no materiality and as they are not reducible to language. The problems of thresholds are indeed more complex to unravel, richer and more ambiguous than the questions of attribution or introjection on which psychology, as we shall see, has focused a great deal. The effects (Threshold effects) are in the domain of sensation, but they are not, however, impossible to grasp. They are translated by actions, impressions, postures and sometimes words. And while they may remain unspeakable, their variations in intensity are often quite palpable in an interaction.

See, for example, how Karan chooses to go beyond the threshold of strangeness (STRANG). He vacillates in his posture between machine-like repetition and an ode to detachment. Repetition can fail to make anything emerge, but it can also turn into indifference, and become a style, making any relationship of trust or empathy complicated. It can also lead to a form of hypnotization. This is the case when Karan calls on his interlocutor to concentrate on her devotion, not to stop praying, like a machine, and tries to bring her into a hypnotic state, which she more or less resists. Being a machine becomes a positive state, a value. Anand is constantly fighting against his own dehumanization, against the coldness and distance that the machine inevitably produces. Modi also uses it in his own way, destabilizing his interlocutor, to whom he predicts all kinds of bad events. The important thing is that going beyond the threshold of strangeness can not only be more or less pushed, but also brings out new possibilities of play.

It would be a mistake to consider that a good divine performance presupposes having successfully crossed each threshold or that one should occupy all the differentiated regions. The question is, rather: what changes when such and such a threshold is crossed, when a peak of strangeness is reached, when an abrupt fall into the sphere of worldly turbulence is felt, or what are the creative implications of having reached one's threshold of critical morphogenesis? Moreover, it is worth thinking about the overlapping or even the superposition of these zones, or the fact that they sometimes combine to increase the amplitude of the catastrophe suffered by the incarnant, or even that it is in fact a single field of disturbance that is progressing and whose dimensions or tones the model makes salient. When Karan appeals to the *Oṃ namaḥ śivāya* mantra, for example, a connection to the cosmos (COSM) seems to occur, but, on the other hand, a sudden disconnection from that of worldly turbulence (TURB) also occurs.

Karan wants to expel himself from the exchange, he fades away while still being there all the same. Karan would probably have preferred to be a ghost at that moment, in other words to take advantage of a repolarization in the spectral field (SPECTR), but he cannot escape completely and remains trapped in the face-to-face exchange despite his desire to disappear. Synchronization with the cosmos fails, but the failure is productive. Indeed, it fails because the bus driver does not follow Karan, he refuses the point of engagement that is offered to him. The failure is productive because Karan, by maintaining that the gods are as impotent as men, that the cosmos does not integrate anything, that it is rather a comfy zone in which to escape with impunity from traffic, pollution, uncontrollable and unmanageable megacities and the monstrous creations of humans, created in total unconsciousness the conditions of his own catastrophe. He finds himself embroiled with his interlocutor in a veritable black hole at the intersection of three zones (COSM + TURB + SPECTR). From the moment he suggests that God can flee from humans, meditation and mantras appear to Nitin as a form of sensory occlusion in the face of the need to act and find concrete solutions to insurmountable problems. God can only emerge reconfigured from such a catastrophic chain of events or disappear. He has reached point zero.

It is likely that every interaction has its zero-point (or points), in those areas of higher turbulence, at the intersection of all zones, where they all meet in a single point of dissolution, where all incarnational strategies fail. And it is often at unexpected moments that such a reset operation occurs. It is when Modi admits his powerlessness as an astrologer to deal with afflictions and confides in his client about his own despair that he reaches ground zero. This precise moment when God and astrology fall apart simultaneously is also a moment of reflection for both participants, who must find a way to sublimate this state.

To feel that we have reached such a critical point is not obvious. We need to pay close attention to the moments when it doesn't work, when the incarnant is forced to stutter, and when the interaction starts up again, despite the apparent failure. These ground zeroes and zero-points are variable in intensity; they can also come back in a cyclical way, occurring in an attenuated form, in a chain, like a computer that one manages to reboot periodically, in safe mode, after repeated bugs, until the final bug is produced.

If Karan had a sensation at ground zero that was qualitatively different from that of his interlocutor and different from that of the spectators of the interaction, we were all nevertheless aware that something had happened at that moment: an intended effect missed its target but did not result in the rejection of the interface, its definitive benching, or else the condemnation of the incarnant. Neither the incarnant, nor his interlocutor, could remain under the impression of an impossible synchronization. A good pact looks to find itself at its very limits, in search of other felicity conditions. For a truly sublime engagement to take place, it was necessary to go through all the failures, for a human to finally tell God how incapable he was, to take him by the hand and explain to him the right way to be a god. And that an incarnation should finally take the trouble 'to inquire a little', as Nitin used to say, even if it means facing the misery of the world in a different way, at the risk of failing in style.

Notes

1. Karel Capek, *The Absolute at Large*, trans. Šarká B. Hrbková (London: Macmillan, 1927 [orig. 1922]), 28.
2. *Ibid.*, 81–82.
3. *Ibid.*, 144.
4. *Ibid.*, 144–45.
5. In *God & Golem, Inc.*, Wiener distinguishes man from machine in the following way: the human brain can deal with fuzziness and vague ideas, a machine cannot deal with chaos, nonsense, fog. Man is exceptionally good at working with the illogical or the irrational to derive usable information. Wiener writes, for example: 'Chief among these advantages would seem to be the ability of the brain to handle vague ideas, as yet imperfectly defined. In

dealing with these, mechanical computers, or at least the mechanical computers of the present day, are very nearly incapable of programming themselves. Yet in poems, in novels, in paintings, the brain seems to find itself able to work very well with material that any computer would have to reject as formless' (73).

6. Barbara Cassin puts the point well regarding paganism at the time of Homer: 'To make the reference point descend amongst the men, the animals and the trees, is initially to multiply it, to pass from God to gods to apprehend a permeability men/gods', because 'anyone encountered could be a god', in Barbara Cassin and Mark Alizart (eds), *Traces du sacré* (Paris: Centre Georges Pompidou, 2008).
7. The way in which humans are led to perceive the person in things or other beings constitutes, simultaneously, a question of psychology, an ethical and political debate and an historical problem. The possibilities of fusion and confusion between object, presence and person (and a fortiori divinity) are, according to the contexts and the specific times, widely exploited or channelled. See, on this point, E. Grimaud, A.-C. Taylor, D. Vidal and T. Dufrêne, 'Qui est là? Présences limites et effets de personne', in *Persona, étrangement humain* (Musée du quai Branly/Actes Sud, 2016), 11–19.
8. Philippe Descola addresses this point in questioning Eduardo Kohn's dismissal of stones from the ecology of mechanisms by which forests think, in the preface to the French edition of Kohn's *How Forests Think. Comment pensent les forêts* (Brussels: Zones sensibles, 2017). Hindu polytheism pays attention to the mineral, as shown by the eminence of stones used for various protective purposes, and as shown by astrology, which makes any plasticity, whether a human face or a geological concretion, the product of wave ballistics. See Caterina Guenzi, *Le discours du destin* (Paris: CNRS, 2013) and E. Grimaud, *L'étrange Encyclopédie du docteur K.* (Nanterre: Société d'ethnologie, 2014).
9. The creatures of the imagination must be 'a semblance of truth sufficient to procure for these shadows of imagination that willing suspension of disbelief for the moment, which constitutes poetic faith'. Samuel Taylor Coleridge, *Biographia Literaria*, 1817.
10. Maurizio Bettini, in *Éloge du polythéisme* (Paris: Les Belles Lettres, 2016), underscores the interest shown in the Greco-Roman era in the gods of others. 'There is a moment when one begins to apprehend the existence of a particular deity and learns to know its characteristics; after which – if it is deemed appropriate – its cult is officially established.' *Éloge du polythéisme* (Paris: Les Belles Lettres, 2016), 102. Wissowa showed, for example, that Apollo was unknown for a long time in Rome before being discovered and integrated into the Roman pantheon. Georg Wissowa, *Religion und Kultus der Römer* (Munich: Beck, 1912).
11. Masahiro Mori, 'Bukimi no tani (The uncanny valley)', *Energy*, no. 7(4) (1970): 33–35.
12. The uncanny has been the subject of very elaborate scales of evaluation in another field, UFO sighting. While working for the US Air Force, J. Allen Hynek, the head of the astronomy department at Northwestern University, developed a scale of 'close encounters' in 1949 to assess the credibility of witness reports of UFOs. See his book *The UFO Experience: A Scientific Inquiry* (Washington: H. Regnery Company, 1972).

13. On this point, see the interview of Mori by Zaven Paré, Chihiro Minato and Emmanuel Grimaud, 'Le Bouddha dans le robot', *Gradhiva*, no. 15 (2012): 162–81.
14. On the marvellous ability of the human brain to attribute behaviours to anything that moves, see Fritz Heider and Marianne Simmel, 'An experimental study of apparent behavior', *American Journal of Psychology*, vol. 57 (1944): 243–59; for an exploration of these mechanisms in the context of robotics, see Valentino Braitenberg, *Vehicles: Experiments in synthetic psychology* (Cambridge, MA: MIT Press, 1984).

11

The Sex Appeal of the Zero-Point

Virtues and Vertigo of Experimentation

What lessons can be drawn from this experiment in anthropology and design? In *Experience and Nature*,[1] John Dewey defines experience as any situation in which an organism enters into an active relationship with its environment, either to maintain its equilibrium or to create a new equilibrium required by its own growth or by modifications in the environment. Experience is therefore to be found everywhere, in all living beings. If experience for Dewey quickly becomes synonymous with culture, to the point that he cannot avoid diluting the notion, in *Art as Experience*[2] he will then try to overcome the undesirable effect of such a conceptual framing. In the latter work, experience is considered as 'the result of interaction between a live creature and some aspect of the world in which he lives',[3] it is diffused, existing in any relation maintained by a body, impregnating technique, communication or art, but it does not, for all that, always have the same integrative virtue.

The major criterion retained by Dewey to discriminate experience from what it is not is the *resistance* of the environment to will or desire. 'The only way it [impulsion] can become aware of its nature and its goal is by obstacles surmounted and means employed; means which are only means from the very beginning are too much one with an impulsion, on a way smoothed and oiled in advance, to permit of consciousness of them,' he writes. And he adds: 'Not without resistance from surroundings would the self become aware of itself; it would have neither feeling nor interest, neither fear nor hope, neither disappointment, nor elation. Mere opposition that completely thwarts, creates irritation and rage. But resistance that calls out thought generates curiosity and solicitous care, and when it is overcome and utilized, eventuates in elation.'[4]

We should not dwell too much on the fact that there is no true experience in Dewey's view except by a 'living being' (and if possible a human), capable of putting to good use what he or she lives, of drawing from it the satisfaction of an acquired knowledge. Nothing prevents us from looking for this aptitude in plants, streams, institutions, spirits, micro-organisms, deities, machines or intelligent systems, since they all have their own ways of being alive, of encountering and overcoming resistance. What matters to us here is not so much to know *who* could be recognized as having this ability – something invisible, for example, in the case of *Ganesh Yourself* – for those who have been deprived of it up to now, as to go down a notch, to the level of the configuration itself, to the *plane of consistency* of the experience, in which there is resistance and the possibility for the being in question to overcome it. There is no experience without a *plane of consistency* endowed with its properties of resistance, with kinetics, with a wind of madness generating an imbalance, an instability, a disturbance or a danger. The idea of an authorless experience, or, rather, one in which its subject is precisely in question, that is to say, the idea of a chaotic reversal of the plane of orientation of the experience, where the one who believed himself to be a subject (such as the incarnant in *Ganesh Yourself*) becomes the object, where the god, the machine and the human exchange their rapports, such an idea opens up, it seems to me, other ways of making, of playing or of experimenting in the field of humanities.

Any current entity, including the most abstract or the most seemingly solid, can, without appearing 'alive' to us, move, liquefy or escape from us, so that it could well be attributed capacities of experimentation with variable intensity. It seems to me that reality becomes particularly rich in problems of all kinds as soon as one starts to think of it in these terms. We could define *experimentation* more generally as the *ways* in which entities play, truly put their capacities to the test, and, forged in contact with new resistances, test themselves on a plane, as well as the ways in which new relationships or new equilibriums thus *precipitate* in provocative configurations.

If it is trivial to say that *experimentation* is everywhere present in this form (preliminary to Dewey's full experience), at all scales of life and not only in human affairs, it is less simple to try to answer the question of what exactly our epoch is the experiment. In the age of the Anthropocene, at a time when the Technosphere seems to have raced out of control, which entities are experimenting with what? If it is legitimate to think that humans are not the only ones experimenting by, through and beyond machines, a host of other entities are at the same time charged with new experimental capacities. How to seize in a particular configuration the sleight of hand by which the one who thought himself subject becomes the object, the experimenter experimented and vice versa?

In *Ganesh Yourself*, humans seek to test limits of various kinds (their own, that of polytheism, that of machines, the threshold necessary for a successful divinity effect), but one can very well reverse the reading of the protocol and imagine that through it, it was a future form of Ganesh that was testing its own viability. Not humans playing God, but a god evaluating the conditions for its new appearance. It must be recognized that outside the normal framework of experimental science, which likes to lock down in advance who is the subject and who is the object,[5] the question 'Who is experimenting with, and experiencing what?' never admits of a simple answer. By further turning the key in the machinery of subjects and objects, let us note that the notion of experimentation maintains a deep affinity with the metaphysical project, if one understands the latter as a refusal to allocate in advance the positions of subject and object (in a sense that evokes what Gilles Deleuze called a 'plane of pure immanence'), and in opposition to any 'transcendental' attitude condemned to fix forever the active and passive poles of the experiment.

Boundary Planes of Experience

If the notion of *experimentation* is certainly useful for qualifying what happened here with *Ganesh Yourself*, it is nevertheless necessary to engage in some serious work to save the term, and not only because it has been confiscated by the experimental sciences. Indeed, in these times of all-out experimentation that we are going through, isn't it enough to observe and patiently report on these innumerable experiments? In the most common-sense use of the term, experimentation designates a particular kind of experience that aims at testing balances of causes and effects that, as long as they have not been tested, remain hypothetical, that is to say, the act of provoking or attempting such balances without being certain of a result. As Claude Bernard wrote in his *Principles of Experimental Medicine*, 'Man can only observe the phenomena that surround him within very restricted limits; the majority naturally escape his senses, and simple observation is not enough for him.'[6]

It is therefore necessary to provoke these phenomena and to create the conditions for their appearance. This is the most common definition of the notion of experimentation in the life sciences, but also in the cognitive sciences and experimental psychology, which readily draw inspiration from it. Of course, we cannot limit ourselves to this usage. Scientists do not have a monopoly on what is experimental, artists can claim that they are experimenting when they venture into improvisation, filmmakers can claim the label of experimental cinema when they go off the beaten track, an astrologer can say that he is experimenting when he tries out other forms of calculation, and supporters of 'intelligent design' will readily say that God is experimenting with forms of equilibrium through evolution. If experimentation thus conceived is constitutive of many of the domains we observe, why has experimentation been locked up in the laboratory when it is as widespread as eating or sleeping and admits of wild or even subliminal forms (situated below the threshold of consciousness) and of which, one day, we will have to take an inventory?

Is it because of the same curious sleight of hand that anthropologists forget the experimental character of their discipline? Are not cosmologies, forms of kinship, systems of exchange and everything that constitutes the diversity of possibilities explored by anthropologists basically experimental forms of organization and classification, that is to say, variants or possibilities of equilibrium tried and tested among other possibilities? When asked why he was interested in sexuality, madness, crime and death, Foucault replied that he had sought to inquire into limit-experiences, historical transformations questioning the limits of humanity.[7] Going from systems of social organization all the way through to rituals, via cosmologies, anthropology has given an account of a whole host of limit-experiences in Foucault's sense, experiences without ascribable beginnings, without announced ends, with no real author and without any other aim than the search for equilibria that are more or less long lasting and sometimes of a miraculous longevity. An experimentation, in the anthropological sense of the term, could be nothing other than *the path taken by a new possibility for opening up a world*; and a limit-experience designates the very imbalance produced in the alliance with the outside, the way of looking for a way out in a new equilibrium, of tipping over or of settling down in an unprecedented covert relation.

In this current historical moment, experimentation is everywhere, overflowing, human beings are both a bit 'experimenter' and very much 'experimented', undergoing large-scale shock waves (ecological, technological) to which they are forced to adapt. What might anthropology look like in this context, one that would be, at last, experimental? This book endeavours to provide an outline of an answer. Anthropologists report on the experiences of others at the same time as having their own, under the influence of an 'evil genius' (*malin génie*) pushing them to pursue a maximal difference, or gap, between the two, roving along on ever more singular pathways. They have not only developed a great passion for distant, marginal objects and limit-experiences without knowing it, they also take a malicious pleasure in building their own boundary planes of experience, that is to say, devices (of observation, of provocation) affording the precipitation of cascading encounters at the very limits, on the *fringes of the unknown*.[8]

The life of an ethnographer is a source (more or less inexhaustible) of events, but not all of these events have the virtue of pushing the limits of our preconceived ideas. How can we conceive of the movement that led, for example, Lewis Morgan to track the ingeniousness of the American beaver, or the strange impulse that pushed Bronislaw Malinowski to take an interest in the sexual games of the Trobrianders, or even the attraction of Lévi-Strauss for mythologies, at the limits of logic? How to conceive of them in terms other than that of a necessity, a necessary attraction to *frontier phenomena*, unlocking possibilities at all costs?[9] Wherever one looks in the history of anthropology, it involves the uninterrupted production of boundary planes, sometimes exaggerating singularities, even fantasizing them, in order to better affirm the field of possibilities. Jean Rouch once confided that what he loved above all was 'putting uncanny objects into circulation':[10] a mission that any anthropologist might like to undertake. Rouch skilfully constructed his plane of disorientation as other anthropologists had done before him. A plane that pushed at the limits of any analytical enterprise, an overflow of images, sounds, movement, setting up direct cinema as the ultimate hope for the regeneration of our concepts, fighting against a drying up of the discipline by way of the transmission of living, raw, barely organized material.[11] What a strange and cunning thing is anthropology: it feeds and strengthens itself only through and in this movement of pushing at the very edges of analysis. The important thing in a boundary plane, and I will come back to this, is the excess of movement, the raw material always exceeding our capacity to conceptualize it. But the notion of a boundary plane would not be of much interest if it applied only to anthropological appropriation.

More broadly, it concerns *any configuration that refers to something that exists at the fringes of perception of an existing entity, or at the limits of its capacities to act, its blind spots.*[12] And I believe that there is much to experiment with in this field, at the time of the Anthropocene, well beyond the God form that has been the preoccupation in this book. Is the Earth not a vast boundary plane populated by blind spots? Isn't a mountaineer immediately confronted with a limit when the plane of the ordinary walk tilts perpendicular, obliging him to resort to a whole set of techniques, ropes and devices? Isn't the surface of the water a boundary plane for a fisherman insofar as he must constantly watch for signals and invent traps that will allow him to catch his prey? Is a forest not a boundary plane when it sends its inhabitants to the limits of what they can perceive, populated by presences whose behaviour they must decipher or anticipate? From microbes to UFOs, from marine life to mineral deposits, from animals to the social life of algorithms, from stars to humanoids, from the most isolated tribe to questions of the climate or geopolitics, there is nothing today that does not call for its boundary plane of experience or that does not generate it in one way or another, putting humans in a truly experimental situation.

This anthropology could well take the path of *Knowledge Through the Abyss* sketched by Henri Michaux. After having declared that 'drugs bore us with their paradises. Let them give us a little knowledge instead', Michaux conceives of a real *boundary plane of experience* within the stupefying encounter in order to question the 'unexpected infinitization' of all that takes place when one ingests mescaline. The finite is caught in the crossfire, overwhelmed, crumbled by 'something of the Infinite, which, far from being a suburb of the gods, is an eternal surpassing ... in any direction, in any object, in any matter, breaking, tearing apart, always beyond, beyond any person however divine one may wish him, beyond no matter how, beyond, beyond, inaccessible, dizzily escaping from any enclave coming from the human mind'.[13]

What could the boundary plane of experience proper to the prison consist in, a situation that is already extreme, or that of a war, or a pandemic, or of what Michaux calls abyssal-situations? Because living with gods, invisible powers, is already a boundary experience, but it is a form of limit-relation to the invisible and to the world that has become banally familiar with time and that may have lost its depth for some. This is where Michaux's notion of infinitude is interesting. I take it up so as to ask what a good boundary plane of experience could be.[14]

There are no rules about how to conceive such an experiment but we only know that ours is a good one because it never stops never-ending, like a book that we never finish rewriting because we want to give it a final point. To have a sense of an ending after having identified the logic of resetting that animates a successful interaction with Bappa (in other words, to have reached what we have called zero-point) is necessarily misleading. The zero-point has its own infinitude (or we should say incompleteness here), and if the questions it raises are abyssal, it seems to me that this is the right time to try to pick them up and see if they might not be useful to others tempted to experiment elsewhere, on other grounds or other objects in the humanities.

Anthropological Design as a Way of Inquiry

There are at least two ways (well, really there are many more) to design a boundary experience and ultimately to operate what is perhaps a *reset* in a universe of concepts and practices.[15] The method we followed here is somewhere between the two. The experimental experiential path that we privileged consisted in getting folded up in an object in the present, which was then endowed with renewed depth. In the same way that Michaux in *Knowledge Through the Abyss* invited his reader by way of drugs to 'ally himself to a dislocating state', we find ourselves 'in metaphysics by way of the organic'.[16] And then there is another way, a properly hallucinatory one, but one vital for resetting the outdated human if we believe the prophets of the Singularity, whom we will come to, who aim to hack themselves. It's less *Knowledge Through the Abyss* than through bugs in the program, even if the second is in fact only a variety of the first, and which would lead to 'anthropology by design'. If the first way is well conceived by Michaux, the second requires more explanation, especially (or all the more so) as we live in a time where technology can be marked with suspicion. I see 'anthropology by design' in my own manner as consisting in finding ways to drill into the concrete at the same time as the abstract, entering zones that could not be reached without the support of a medium, a device, a contraption, a collective experiment or even an *anti-machine* as we did.[17]

The history of speculative design, this branch of design that aims at making quirky offbeat objects to provoke debate, new practices or a reflexivity about how objects are used, which Bappa certainly points to, is even shorter than that of the philosophy of machines. Let's simply say that the machines recognized for their philosophical virtues are less numerous than the philosophers who have taken up the theme of machines. Many have seen in artificial intelligence the opportunity to conduct thought experiments and to rethink the philosophy of mind.[18] Few, on the other hand, have ventured to design machines, devices or UFOs in the same way that speculative design does today. However, some machines, and well before Turing's computer, have acquired a singular aura, and have been promoted to the rank of real metaphysical artefacts, shaking up the frontier between science and art, making the human undergo an abyssal troubling.[19]

Anthropology by design has a bright future ahead of it, and not only because more and more anthropologists have realised the potential of prototyping,[20] designing games or contraptions in their own research, to rewire problems and questions; but, rather, because there is sometimes no other way to deal with a problem than the design of devices with strange deforming mirror effects, there less to help or serve us but, rather, to radically question us about our condition. It's their existential virtue. It seems to me that by way of machines, humans have found themselves on the very frontier of themselves: to do more than just ameliorate life, or make the world better to live in, but in order to scare themselves, to put themselves in danger, to touch the zero-point of their own reboot. If it is true that by scaring ourselves we also feel more alive, the richness of the experimentations undertaken in this domain today is such that I cannot explain it otherwise. Just as in Vedic times, when ritual and sacrifice was used to do concrete metaphysics and to experiment with *entities*, in the second half of the twentieth century, after having buried God and then Man (dixit Nietzsche, Foucault and Deleuze), we have taken up residence at the post-human frontier, cultivating the most varied artificial intelligences; including, and especially, at those moments when our world was felt to be in danger. Artificial intelligence is the only field in which investments have been massive after each catastrophe since the Second World War.[21] There is undoubtedly something here that cannot be explained purely and simply through economic reasons. There are also elements of faith or what William James called belief in the world, fundamental anthropological and metaphysical concerns, for which the post-human frontier has become the privileged playground.

As evidence of this, once the Mumbai experiment was over, Bappa continued his life, travelling outside India, staying in various universities and museums, attracting philosophers, doomsday prophets, engineers, tinkerers, artists of all stripes, wondering about the future of humanity, that of machines and also, if a little less often, that of gods and elephants. It was exhibited in Dresden at the Polytechnic Museum in an exhibition entitled *When Machines Are Dreaming* (2017) while the film that we had shot during the experiment, *Ganesh Yourself* (Arte, 2016), went from festival to festival. In the gallery of the museum in Dresden, surrounded by German hackers, artists, engineers and cyberpunk prophets who had designed machines no less bizarre than ours, we felt as if we were at the exact point (a zero-point?) where Gilles Deleuze had warned us that science and art would be, thirty years earlier, surrounded by experiments involving 'the ensemble of a human condition, or post-human, suprahuman or superhuman'.[22] Heading up a collective of hackers, Andreas Ullrich had gathered a whole series of disturbing machines, resulting from the third generation of the cyberpunk movement. *When Machines Are Dreaming* was meant to reflect the experimentations of our time. Gone are Turing and the machines that try to trick their world. No more humanoids or anthropomorphic machines. All that has become outdated. It was necessary that one could imagine unthinkable machines, in the process of being conceived. And that the machines come to terms with themselves as such, that they display their electronic circuits like *écorchés* (anatomical figures). Above all, these machines share a very strange particularity. They could start *dreaming*, making use of algorithms that allow them to feed on images emanating from the network in a random mode. All this under the auspicious gaze of our Ganesh, meditating.

Thinking With, Against and Beyond 'Machines'

Deleuze was certainly a visionary. But how to go beyond *A Thousand Plateaus* and its overdose of machines of all kinds, circuits, coding systems, liberating, desiring, abstract, organic or cosmic machines? Before *A Thousand Plateaus*, Lewis Mumford, in *The Myth of the Machine*,[23] delighted in demonstrating the extent to which the great ancient civilizations, from Mesopotamia to Egypt, had shown terrifying ingenuity, conceiving amazing machineries in order to make their works possible, machines essentially made of humans envisaged as cogs of immense mechanisms. The technological wild takes unexpected forms in Mumford's work: rituals are conceived as great machines, the administrations of great empires become the privileged sites for the conception of monstrous machines, real or imaginary (Chinese celestial bureaucracy, Hobbes's Leviathan and so on.). This freedom from the narrow (technocentric) definition of machines has been succeeded, it seems to me, by another age of thought that is far less exciting, where, curiously, the concept of a machine is now applied only to explicitly technological objects, computers and robots, rather than to all those untraceable things, things that are both invisible and real, abstract mechanisms at the crossroads of the organic and the political that come to machinate beings.[24] But the point that appeals to me most in Deleuze's work is not so much the fact that machines are present everywhere in *A Thousand Plateaus* and disseminated in sometimes unthinkable or aberrant forms. It is the faith he expresses in what he calls machines of the third kind,[25] referred to in his book on Michel Foucault. In the transcript of one of his lectures on Foucault, from 25 March 1986,[26] Deleuze ventures something that would commonly be called a *prophecy*:

> And, once again, what we can affirm in all confidence is that the form 'God', the form 'human' and the form 'superman' exhaust absolutely nothing, they do not exhaust the universe of forces and, certainly, there are other relations of forces between the forces in man and the

> forces of the outside that give a place to wholly different forms which are neither God, nor human, nor superhuman.

And in this same lecture, it is towards India that Deleuze invites us to look:

> I believe that there will be, that one could find, forms which in effect will be no longer God, nor human, nor the superhuman, but which will be extraordinary vegetal or animal forms ... It is India that interests me above all in this regard.[27]

Further on, Deleuze shows the same excitement, undoubtedly tinged with fear, with regard to this new horizon that he perceived well before the transhumanists inundated us with apocalyptic techno-prophecies:

> The machines of the third kind represent a veritable revenge of silicon on carbon. And now it is silicon that presents the incontestable advantages. One can conceive machines of the third kind on carbon. Moreover, it has even already been announced, for the computers of the future, the machines of the future to come, will return to carbon. We are indeed told that the true robots will be organic robots. There will be ... this is a bit of a detour, it would be beautiful to give it a figure, we are in full prophecy mode.[28]

Deleuze then adds,

> the relation with silicon is incarnated in certain technologies, but it overflows technology a great deal. The history of the use of silicon, etc., infinitely overflows science since it will animate forms of art, it will animate all sorts of things, everything that puts in play the whole of the human condition, or the post-human, suprahuman or superhuman condition ... whatever you want to call it.[29]

Deleuze saw in the post-human a real horizon of renewal, an opportunity of formidable creativity, an opportunity of mutation, well before the explosion of the paltry transhumanist prophets, the pop philosophers, writers of science fiction and futurologists who see themselves today as the metaphysicians of the century to come. It is necessary to recognize that the impulse of creativity announced by Deleuze really did take place. Under its mutant forms, as much real as metaphorical, the machine today haunts us, it invades us, it has penetrated our modes of thought, it has intruded into debated domains (sexuality, war etc.), much more deeply than we could have imagined. It is not enough to note that

Deleuze and Guattari had predicted it, making an act of philosophical exorcism by passing all our relations through the screen of desiring machines, nor to recall that Philip K. Dick, had also prophesied it but in a much more dystopian way, inventing in his novels the most anxiety-provoking cybernetic machines.

The universe of machines is even more vertiginous today than it was yesterday, conducive to amazing metaphysical experiments. It is unnecessary to inventory them here. The most intriguing machines, arts and sciences included, are those which produce a real experience of vertigo, facing humans with their smallness, their uncertainties, the finitude of human nature, the clumsiness and the weirdness of our cognitive mechanisms. And those who are tempted to resist the empire of the machines or to question it have no other choice than to try to subvert it from the inside or to act in these margins, forced to invent boundary planes of experience and even anti-machines as we did, which question the machine experience. *Acting and thinking with, but also against and beyond machines* that can take very diverse and sometimes unsuspected forms (metaphorical or invisible) is probably one of the great collective experiments of the start of the twenty-first century, in which we are all, whether we like it or not, stakeholders.

Philip K. Dick, in a text entitled 'The Android and the Human', a few years before Deleuze, himself observed the vertigo produced by the machines of his time, while also making himself into a bit of a prophet. It is from the angle of the attribution of intentionality that he approaches the problem, however, rather than from that of the dissolution of the human and its recomposition in connection with other forces which, if it greatly fascinated Deleuze, in no way delighted the worried humanist that Dick was, one nourished in Christian mysticism.[30] Preoccupied by the fact that the environment is evermore imbued with technology, machines and animacy in an artificial mode, he wished to draw attention to the fact that the world is populating itself with vague, confused entities, entities favourable to awaking doubt as to their nature, suitable for provoking a metaphysical revolution whose importance we do not yet gauge. The difficulty, according to Dick, is to know how to avoid the reduction of this complex problem of civilization to the question of psychological order that we usually call the 'imputation of intentionality', a bad concept, according to him, because it softens or strongly attenuates the metaphysical potential of machines, that is to say, their capacity to call into question categorical dualities (body and soul, living and non-living), but also some well-anchored anthropological certainties, such as human superiority, its singularity or its intelligence.

Starting from the idea that there is 'the tendency of the so-called primitive mind to animate its environment', he insists on the very curious role played by modern psychology which 'has requested us for years to withdraw these anthropomorphic projections from what is actually inanimate reality, to introject – that is, to bring back into our own heads – the living quality which we, in ignorance, cast out onto the inert things surrounding us'. The introjection would be then the 'authentic mark of civilization' which would distinguish us from the primitives who see their natural environment as 'pulsing with a purpose, a life'. The mature, scientific and sane individual would thus be condemned to eliminate these 'childish projections' in the name of the principle that 'the world is dead, and that life resides solely within himself'.

'But one wonders', Dick ironizes, 'has he not also, in this process, reified – that is, made into a thing – other people? Stones and rocks and trees may now be inanimate for him, but what about his friends? Has he not now made them into stones, too?'

And Dick is not content to denounce the role played by modern psychology in convincing us that everything is under control and that in our relationship with machines there is no enigma; on the contrary, he points to the blind spot, seeing in it what Michaux would have called an *abyssal-situation* to be unfolded:

> within the last decade, we have seen a trend not anticipated by our earnest psychologists – or by anyone else – which dwarfs that issue: our environment, and I mean our man made world of machines, artificial constructs, computers, electronic systems, interlinking homeostatic components – all this is in fact beginning more and more to possess what the earnest psychologists fear the primitive sees in his environment: animation.[31]

It would be easy to demonstrate that animation constitutes for Dick a form of the *zero-point*. Each time that an intelligent otherness reveals itself, that an entity coming from elsewhere unveils its properties, it is in the form of an unimaginable or inconceivable movement that throws our modes of apprehension into disarray. And Dick underlines one of the major implications of this revelation, and especially the centrality that the concepts of *behaviour* and *analogy* will play, from now on, in this world that is being reanimated (in the quasi-medical sense of the term): 'it is now possible that we can learn about the artificial external environment around us, how it behaves, why, what it is up to, by analogizing from what we know about ourselves'.

It is interesting to note that at the same time that Dick was criticizing psychology as incapable of dealing with the metaphysical conundrums posed by machines and inviting the invention of an experimental theology by marrying Turing, St Augustine and Meister Eckhart in borderline science fiction scenarios featuring superhuman entities, the idea that one could have a psychically deranged relationship with non-humans in general, and with machines in particular, was being explored in some strands of psychiatry. Harold Searles, for example, had noticed that many of his schizophrenic patients, at a relatively advanced stage of the illness, saw themselves as machines or under the influence of uncontrollable mechanisms. They could also see themselves as animals or plants, but the anguish was never as intense as when they felt that they were a mass of circuits and bolts or that they were under the control of a machine outside of themselves that was dispossessing them of their sensations or taking over their vital functions. Searles proposed the term 'relatedness', the intuition of an intimate relation that the individual would share with the non-human elements that surround him (atoms, molecules, metabolisms, behaviours), as well as the maintenance of a 'reasonable' relationship that incites him to become aware of his singularity, knowing that the quest for fusion would mark the disappearance of his individuality.

We know to what extent Dick enjoyed frightening his reader in the mode of 'My God, what if ...', confronting readers with forms of 'unreasonable' relations, unreasonable according to Searles's criterion, where mergers and confusions of personality multiply between humans and machines endowed with unstable behaviours bordering on cyclothymia. Nobody was more conscious of the fact that, at the time of the first computers, we were already involved in a form of collective limit-experience of which we could foresee neither the end nor the bottom, and that science fiction was in the process of becoming the 'popular metaphysics of our time', as Viveiros de Castro so rightly put it.[32] A very particular metaphysics, which feeds on extrapolations and projections into the future as thought experiments. What exactly is the post-human frontier if not the site of a mass of experiments with limit-experiences, sometimes partial, sometimes radical, on aspects as diverse as memory, thought, sensibility, relations, communication? There is not a single area of life that machines have not infiltrated, causing a radical rethink of things ranging from sex, to war, and politics. A strange tabula rasa effect is produced, whose intrigue is just getting going.

Finding Our Way Out of Techno-Prophecies

Let's go back to *When Machines Are Dreaming*, the exhibition in Dresden at which hackers from all over the world met up. All the machines exhibited there seemed to belong to the same subversive or reflexive lineage, *thought experiments* concretized in objects. Imagine Bappa sat enthroned, eyes half-closed, with meditative countenance, in the middle of an exuberant zoo of sometimes poetic, often useless machines, experiments conceived by programmers for whom dreaming is more important than intelligence. Bappa sometimes opens his eyes, only to close them again immediately, in a loop. Philip K. Dick would probably have had a lot of fun with the highly speculative possibility that machines might not only think but dream, explored by Ullrich and his friends, who were already at that time determined to make hallucination an operative concept to talk about the inner life of machines.

Option 1: Machines do not dream, but they will one day, soon.
Option 2: Machines already dream, but we just don't know it.
Option 3: Machines dream intermittently and sometimes go into altered states of consciousness that are unknown to us, etc.

By dint of having reduced the philosophical problem of machines to a question of intelligence, we have forgotten all these unclassifiable states, all these ways of being a machine, in automatic or random modes, which cannot be assimilated to dreaming or waking, or even to sleep. This sums up the kind of curious zone, between the speculative possibility and the very real unknown, that one begins to explore when one takes the path of machines. It is confusing, clumsy, often out of step with real technical possibilities, but it helps to imagine an otherwise inconceivable possibility.

By contrast, the program for Ganesh is poor, low tech, unpretentious, but surprises the visitors, who willingly stay to contemplate the synchronization between the eyes and the random movement of the trunk.

> Eyes open at one minute thirty. Contraction of the trunk at one minute thirty-eight. Eyes close at one minute forty. The eyes open again, this time synchronous with the trunk, which perhaps translates a gastric spasm at two minutes and five seconds.

Who is responding to whom, with the eyes or with the trunk? A real spectacle, with minimal means. The system is self-sufficient, a simple presence. Or pure absence, because the person whose recorded face is now in the Bappa robot did not meditate for very long, he only pretended. His simulation, however, takes on another dimension, incorporated into the machine. For the visitor, there is no doubt: Ganesh meditates. It is difficult with this kind of machine to escape the poetry of 'as if' and not to let one's imagination run wild. Andreas Ullrich said to me that perhaps 'Ganesh is getting imbued with his mission, [that] the machine is learning meditation', or that God is learning simply *to be* in a silicon body, without speaking. Far from the crowds, the effect of presence is finally apprehensible in itself, as the product of an original emptiness that is self-generating. If *Ganesh Yourself* was in part an experiment and experience in layers of depth for both the incarnant as well as for the interlocutor, it was that it suddenly allowed the participants to just barely touch this emptiness, this fragile weightlessness, from which the divine can be felt but also can be undone just as quickly.

In the gallery in Dresden, an unexpected synchronization took place between Ganesh and the other machines that surrounded it, machines much more sophisticated than Ganesh, which were plugged right into the network. Ganesh's trunk is subject to jolts (bugs?) that are impossible to predict. It is equipped with sensors that are supposed to react to light, but in Mumbai we could not always explain its movements, as if the trunk sometimes reacted to speech or to the state of excitement of its incarnant. In Dresden, *something* seemed to be going on between the machines, and with respect to this something, to our great surprise, it was only Ganesh that embodied it, he personified it with his face, when his eyelids lifted or closed or when his trunk gave the impression that it was gesticulating in response to other machines. A few machines generating an impression of synchronicity, even if not really connected, is enough to create the idea of an electronic (taxonomic) kingdom. Paying attention to Bappa's every twitch, Ullrich then tells me, 'There's no reason not to think that there are real synchronicities between all these devices.' If Ullrich is to be believed, machines could well be pushed onto supernatural paths, and unknown forms of superhumanity or supermachinery hatched in hacker garages. It's hard to imagine, but anyone who wants to touch the living possibility, to feel it in their flesh, would have benefited from a stroll through the museum in Dresden that day.

In a curious effect of convergence, hackers, engineers and prophets share the same eschatological motives today. The kind of science fiction produced in counterculture circles at the end of the twentieth century has become the dominant ideology of Silicon Valley in the twenty-first. Like others, Ullrich believes that machines are the future stage of life, that humans will become extinct like dinosaurs, and that new cybernetic entities now in the making will eventually absorb the psychic abilities of humans. 'This will be the end of human bodies as we know them', he told me, 'not of the Earth, but of the human, and the advent of a new life form.'

Put another way, Ullrich was pointing to nothing less than the Technological Singularity,[33] a cataclysm comparable to the Cambrian explosion of life, from which our species should emerge entirely alien to what it is today and from which an incredible proliferation of artificial life should emerge. How to explain that even Ullrich, this German hacker and convinced ecologist, family man who grows delicious 'home-made' herbs in his garden, has been so caught up in the prophecy of the Singularity? The post-human frontier appears as a kind of final limit, the ultimate horizon of tinkering and devotion for *Homo sapiens*. Promising new sensations, it gives him a direction and renews the relationship to the world of post-industrial societies. Everyone is obliged to position themselves in relation to this frontier in one way or another, either by refusing the invasion of machines, or by experimenting on its margins, or by seeking other forms of humanism.

In this critical, experimental and dystopian moment, Ullrich, his hacker friends and cyberpunk artists, made their choice. Not to refuse the machines, but to subvert them. So they were busy fighting for a world of more human algorithms against other algorithms that were deemed inhuman. According to them, there was no other way out than to engage in a frantic struggle for the mastery of technological environments, even if it meant looking elsewhere for other gods or other forms of devotion, in order to preserve one's humanity or to be otherwise human.

It is not surprising, in this prophetic climate, that Bappa finds itself in Dresden and attracts the attention of pirates, clairvoyants and techno-prophets, like flies haunted by the spectre of what might come. It is true that this little machine of divine deconstruction reached a form of apotheosis when environmental activists seized upon it to promote it as a recyclable idol and protest against the pollution of plaster idols. One cannot play at incarnating God in a tele-presence interface without stirring up hope or fear, taking on all the world's problems, launching prophetic appeals, stirring up millenarian fears or raising ecological alarm bells. Jacques Ellul would certainly have considered all this as the symptom of being under the sway, or a symptom of *possession*, whose precursory signs had long been visible, when, by a strange magic operation, the sacred was transferred to the technical.[34]

Even so, I'm surprised that Bappa could be considered an alternative to anything: it was not designed to be worshipped but to make people think. This metaphysical gizmo introduced into the blind spot of the beyond was meant to reset God or, more humbly, the mechanism of devotion, not the human. But those for whom machines have a world (*Umwelt*), one we don't know about, those who work for the slow awakening of electronic life or who campaign to 'liberate' computing, see things quite differently. It is through contact with the machine itself that we will touch the zero-point of the human being, that we will find the means to carry out its reboot, without any other God to adore for the moment than our good old monotheistic god, the one from Norbert Wiener's *God & Golem*,[35] or Teilhard de Chardin's 'Noosphere' and 'Omega Point',[36] notions all equally imbued with monotheism, or even computing what some already describe as *celestial*[37] and which will only reveal its true face as a goddess on the day when computers become one with *Digital Gaia*,[38] an Earth riddled with sensors and information.

Faced with these techno-prophecies, one can always argue that other contracts are possible, that other ways of deifying and personifying the components of our environment have already been tried, and are yet to be invented, against those who would like to impose their own way too quickly. But prophecies are not forms of anticipation like any other. They impregnate us with potentialities despite their apparent impossibility. By means of disaster scenarios, they tell us that what is important is played out here and now. 'If the Singularity doesn't happen, let's make it happen!' say the Singularitarians.[39] The ones that sound the most frightening are often the most vivid hypotheses in William James's sense, because they got to the heart of History, in an imminent temporality – in other words, to the heart of the zone of worldly turbulence, the zero-point or a reset effect that was identified as being the very mainspring of a successful experiment with Bappa. But with Bappa, a reset does happen, astonishingly, before our very eyes. No one expects it, no one causes it. It happens.

A Thingamabob in the Blind Spot of the Beyond

Poaching on a deity's land, Bappa produced its effects, and sometimes even sparks of wisdom, in those who played the game of 'hacking', or pirating, which consisted in putting themselves in God's place. But our goal was by no means to build an artificial intelligence or a Supreme Being, like some of the mystical engineering projects in Silicon Valley that have emerged in recent years and that are not only dangerously devoid of humour and reflexivity but are also modelled on a preconceived, monotheistic and theologically impoverished idea of what to put under that intelligence.

Our project was more modest, to start by playing at making a god rather than making a god, even if in the universe where we are situated (let us remember the words of the ritual officiants) the result is the same and ends up making a *substitute*. If Bappa could not really claim to be clairvoyant, since it was devoid of an intelligence of its own, it nevertheless made it possible to reinitialize all that God possessed in terms of contained power, to unravel it in a different way and to open up other possibilities, other relations to the magma of invisible forces. The adventure was perilous. It was a question of devising the device where it can materialize only at the risk of imposture. We could have been satisfied with a voice, but we chose to inject a little more body through the image of the incarnant in an interface that did not necessarily need it and to limit the robotic element to a minimum, enough to keep the *feeling* of it. Perhaps it was necessary to start this way, to take up artificial intelligence back to front, to 'tele-operate' God first, to keep it under control, instead of aiming for autonomy. Bappa proved not only to be a good metaphysical toy encouraging speculation. Some have used it, to our surprise, to make a radical critique of the contract of ordinary religiosity, turning the interaction into a court of law, subjecting the incarnant to all sorts of trials, questioning any possibility of representation or the efficacy of God itself. The device of producing an unstable spectral glow, by way of an overhead projector, as fragile as it is capricious, allowed us a gain in sensitivity on a few crucial subjects and to point out threshold and engagement effects (*effets d'embrayage*) in matters of incarnation to which we had previously paid only a little attention.

An experiment without a veritable originator, *Ganesh Yourself*, once launched, renewed itself with a minimum of intervention, precipitating within its set-up people, things, thoughts, even levels of reality. Along the way, those who thought they were appropriating the machine to lure their fellow human beings found themselves lured and those who thought they were engineering the voice of God were engineered by God itself. As for those who thought they had invented something, it is needless to point out that they did not in fact invent anything at all, they had simply given another opportunity for incarnation to a matrix of forms ready to actualize itself at the slightest opportunity offered to it. These reversals are inevitable when one plays with such entities; recalcitrant, unstable, evasive, they always emerge as winners, strengthened, more alive than ever. This is probably a lesson to be learned from this experience. *The effects of superhumanity are generated with all the more ease as they involve reversals of position.* 'Give us a machine, if we can't make a god out of it, the invisible itself will take care of it!' The ease with which the recirculation took place should not, however, make us forget the essential point: the incarnants and the interlocutors never ceased to doubt, to assert their scepticism or their freedom of thought in the face of the proposal made to them. The machine of divine deconstruction was not adopted naively by devotees ready to accept everything. It found its way into their midst and by their consent.

This is because Bappa was not a god. He was in fact both an *anti-machine* and an *anti-god*, a reflexive twist, machine enough to make himself a representative of the kingdom, but not sophisticated enough to claim more, without the complicity of a human being. A mass of delicately stretched wires and analogue cables, it had its own way of being from another age or of producing resistance, by its fragility. Ganesh's trunk was held by only a wire to his sensor, and it was necessary to take care to reposition the counterweight regularly or else the trunk wire would get tangled. Fragility was at its max, the mechanics and the program kept to a strict minimum. As for its difference from a classical idol, it was obvious: turning the mirror towards the devotee, Bappa was nothing more than a deification impulse which seeks itself, crystallized in an object, and not the other way around. He was therefore an *anti-god*. Who could have believed that it was possible to do politics and even metaphysics with such a thing? Very few things were needed such that quests for clairvoyance, as well as the most animated hopes, including the hope that the Occident, rightly or wrongly, put in machines, were to meet so as to produce introspection.

Indeed, once the hope that the Occident puts in machines has been externalized in hyperorganisms or in various artificial powers supposed to extend our capacities, the problem that Western societies have faced is that it is difficult to reintegrate positively or to reincorporate into being its machinic dimension. Prior to machines, it was a problem which also arose with the gods, as we saw with Ganesh. How to reincorporate the divine power once externalized? We have seen to what extent this preoccupation animated our incarnants and their interlocutors, in search of new conditions of bliss. It was therefore necessary to ask the question of what robotics and artificial intelligence would have become if we had completely changed their experimental basis and if, instead of the Turing test based on a competition between man and machine and an apologia for a successful ontological confusion,[40] other tests had been conceived, cultivating poetic faith in fuzzy entities.

When approaching Bappa, one does not say to oneself: 'Another god to adore!' but, rather, 'What benefit will I be able to gain from this curious thing?' or 'What could I say if I were a god?' In other words, we don't expect miracles, but we cultivate an attitude of receptiveness to the context in which divinity could express itself, whatever the support that is facing us. Whilst making it possible for individuals to reorient themselves around the divinity which is in them, this small machine of divine deconstruction fractures being, fractures its own organization at the same time as that of the god which it incarnates. Let's remember all those moments when the incarnant, in order to get through an interaction with an interlocutor who was a bit aggressive, transforms himself into a machine reciting formulas (mantras), justifying indifference or lack of empathy for the interlocutor by the fact that God could well be only a mega-machine, the first ever invented.

Whatever the tactic adopted by the incarnant, the interaction becomes the plane of consistency where fractures and reorganizations take place, where the god falls from his pedestal, where the human being loses the faculty of reasoning and where the machinic aspect permeating the god and the human shows itself at the height of its possibilities of prostitution. *Ganesh Yourself* is ultimately a positive technique of dehumanization, to find the *non-human* in oneself with all the ambiguity that the breach comprises once opened (towards the more than human, the less than human or the downright other). Once the split had been made between his divine being and his machine being, the incarnant has no other choice than to recompose himself in a different way, through these component pieces, to find his line starting from these cracks, to invent *his* Ganesh, or else one that emerges through the intermediary of the incarnant, a singular Ganesh without equivalent in the pantheon of the existing gods, at the same time as a new figuration of the incarnant's self. Without needing to replace nerves with electric wires and senses with sensors, to call this form 'human' might not quite be right, as Nietzsche had prophesied, simply because he will have composed himself differently.

The Zero-point as Technique of Regeneration

Take a step back. Why did we want to go back to zero? We didn't. Our original scenario was more modest; it was the participants in *Ganesh Yourself* who made it a *vital* question, pushing the interface to the point where the god is made and unmade. We could grasp it only afterwards: Bappa's function was to reboot; a reset not only of God, of man, of devotion, but of everything that, under these notions, coagulates in the circuit, becoming so solid as to be fixed, and that is possible to put back in motion via the interface. We would never have imagined that such a deconstruction would take place, and above all that it would be done with such virulence, and even sometimes with malicious pleasure on the part of the interlocutor faced with a failing and flailing incarnant. This enjoyment (*jouissance*) irrupting as the zero-point approached, that is to say at the ultimate point of deconstruction, revealed itself as such only with use, but it is this enjoyment that constitutes, for me, the real enigma. Why such relentless effort to dissolve God only in order to better recast him, to destroy him in order to better save him?

Incarnation after incarnation, Bappa slowly gained his status as anti-god or anti-machine. It was not given, and in no way was it a project imposed from the outside, as if reflexivity was only on our side of the experiment, and that on the other sides there was only ignorance and naivety. On the contrary, all the questioning, the creativity and the intelligence that we witnessed were on the side of the participants. As for the pleasure that each participant could feel in putting God back to zero, it was most often timid at the beginning before manifesting itself with more and more virulence. And no one let anyone dictate where to place the cursor between playing at making a god and undoing it to test what power it really contains. This is not to say that there was not something in the device itself particularly conducive to the *sex appeal* of ground zero, but one can point to it only in retrospect, after it had met its audience.

One could surely imagine other *ex-centric*[41] devices allowing one to put oneself in the place of anything or anyone; in the place of animals, plants, of various entities, visible or invisible. Our troubled times require us to indulge in such play with perspective, in order to gain sensitivity, empathy, to experience the point of view of radical strangeness. But what the experiment demonstrates is that, taken as excentric devices, gods and machines are not just anyone, or anything, and that the point of view of anything, including a god, is not won so simply, not without a dynamic test of radical dissolution. The machine indeed allows Bappa to be a 'god outside of himself',[42] but, by forcing him to interact and enter into dialogue, he finds himself attracted to, and even caught up in, the ordeal of his reboot. In the same way that gods unleash passions, the desire to adore them or to bring them down from their pedestal, robotic machines also produce excentric otherness and, although they have been designed by humans for humans, unleash technophobic impulses, the desire to destroy them, violent forms of rejection or fear, even though it is enough to simply unplug them. If Bappa, with his elephantine air, allows the machine to be 'out of itself' (or else, to be a little bit more than itself), the desire to destroy some god or to make a radical criticism of it meets here the thousand and one ways for a machine to malfunction and to be reduced to a crude support.

It seems to me that the *sex appeal* of the zero-point concerns not just our little experiment. It is also essential for understanding both the incredible proliferation of prophecies of dissolution, like those we have explored that populate the frontiers of the technosphere (Singularity etc.), or even the calls for a *reset*, which are regularly expressed in modern ecological thought, in order to radically reform massive entities (hyper objects, as Timothy Morton calls them), such as modernity, capitalism and the Anthropocene.[43] Is it not risky to link dissolution, taken as the grounds of a prophecy, to the regenerative function of the reset that we observed here, on a reduced scale so to speak, within the limits of an interaction? A philosophical explanation should necessarily take into account the attraction that the possibility of experiencing in real time, here and now, the vertigo of a reset can exert on the human being, beyond the fear aroused by its own mutation. Without it, no regeneration is possible.[44] Let us remember what Zarathustra said: 'The human is a rope, fastened between the animal and the superhuman – a rope across an abyss.' To live in this context would bring about vertigo, a limit-experience, living beings being forced to evolve at the limit of themselves, to experiment or to play, in a state of permanent imbalance, even if it means artificially provoking this state when it is not naturally caused. For the participants of *Ganesh Yourself*, there was nothing abnormal in the fact that God, as a living entity, submits to the same movement of regeneration at the limit of itself, even within the limits of a theatre like that of the incarnation game.

Terrence Deacon says precisely this in *Incomplete Nature*[45] when he writes of the driving efficacy of *absence* at all scales of life, of a constitutive incompleteness, of an irresistible force exerted on very varied living beings (cells, bacteria, plants, minerals etc.), producing an unconscious attraction to an outside, to get out of themselves, and which makes it such that there is creativity, emergence, novelty. In the case of *Ganesh Yourself*, I think that the desire to interact and to make the interaction often last until its breaking point is explained less by a taste for exacerbated play, a destructive drive or a political will for reconstruction, than by a kind of impulse or metaphysical attraction, the importance of which one rarely suspects in social relationships, a taste for depths, a will to cross the threshold of the meta, which has taken very particular forms here, often intensified. Michaux writes: 'As much as with sex, and more singularly, God is released in madness. The immanent god, the experimental god. It will no longer be let go of.'[46]

The *sex appeal* of the zero-point is a drive-force. One does not succumb to it for nothing and it is not actualized just for the pleasure of the game. A real reset is not something to be taken lightly, it has consequences, real implications when we are talking about something as important as God and a fortiori about other objects likely to undergo the same operation of fragmentation.[47] We could tackle other enigmas, with other devices. Is it necessary to list them? Certainly not. Boundary planes of experience, whether they are those that we make for ourselves with meticulousness as well as those in which we find ourselves inevitably embarked by the force of things, can coagulate from any environment, site, interaction or situation. One cannot presume in advance their principles of genesis and renewal. They are essentially made of blind spots, blind spots but also of thresholds of clairvoyance, sometimes within our reach, sometimes impassable. Dynamic configurations, they are measured by the volume of disturbing perceptions and enigmas that they generate. We must learn from this observation about creativity in science and art, about the philosophy of experience in general, but also about the way anthropology can take up and take on the experimental aspect of its practice. One thing is certain: it is no longer just a matter of the humanities patiently inventorying the possibilities that people have of living differently than they do and repopulating the world with all sorts of populations, minorities, divergent interests, categories of beings, visible or invisible. The humanities today have no other choice than to renew their modes of investigation to better define the fractured landscapes that surround us, they encounter, whether we like it or not, multiple 'uncanny valleys' in Mori's sense; everywhere they face the abysses of which Nietzsche spoke, troubled zones, abyssal-situations, as Michaux called them.

Boundary planes, thus conceived, are everywhere when one examines the Earth and when the human being gets involved, at the frontier between ecology and technology, beyond the borders of taxonomic kingdoms, those of the living and the non-living, the animate and the inanimate, the visible and the invisible. It is then necessary to experiment, to survey these unknown regions in a more active way, to point out the blind spots of our points of view, starting with one's own, to isolate effects, to identify critical thresholds, in short to have a truly living vision of these zones. It is only by experimenting that we can catch a glimpse of what blossoms at the limits of perception, to finally be attentive to what we no longer see, in the hollows and interstices, at the intersection of entities present in the situation. It is necessary to progress here with as much prudence and humility as Mori does in the 'uncanny valley', because the valley is constituted from a blind spot of feeling, and it is much easier to try to mimic the point of view of another than to be attentive to what is ignored or not detected in the blind spots. Heavy metaphysical problems could well be granted in the operation of another depth. If the experimentation is productive, it is to the extent that it permits the consideration of seismic movements, in the liveliness of the action, that would otherwise be imperceptible. These have perhaps more often than we think a 'pure center at the same time as being incredibly displaced, as in the focal points of multiple ellipses' (Michaux), a zero-point. Is it not at the very moment of their dissolution that we realize most acutely what matters, whether it is our gods, our convictions, our habits or the very contours of our humanity? Whether the disappearance is real or simulated is of little importance as long as it allows us to apprehend what previously escaped us. Touching the zero-point, everything is renewed: the virtues and vertigo of a *reset*.

Notes

1. John Dewey, *Experience and Nature* (Chicago: Open Court, 1925).
2. John Dewey, *Art as Experience* (New York: Perigee, 1980 [1934]).
3. *Ibid.*, 43–44.
4. *Ibid.*, 59–60.
5. There are, however, many breaches of this principle; see, on this point, Katrin Soldhju's work on the notion of self-experiment (*Selbstexperiment*), in particular on Gustave Moreau de Tours' self-experimentation with hashish, *Selbstexperimente. Die Suche nach der Innenperspektive und ihre epistemologischen Folgen* (Berlin: Trajekte, 2012); as well as the work of philosopher Vinciane Despret, who has uncovered numerous ways in which animals impose their own plane of consistency on experimenters, from *Hans le cheval qui savait compter* (Paris: Les Empêcheurs de penser en rond, 2004) as well as *Habiter en oiseau* (Paris: Actes Sud, 2019).
6. Claude Bernard, *Principes de médecine expérimentale* (Paris: Presses universitaires de France, 2008 [1877]).

7. Michel Foucault, 'Entretien avec Michel Foucault', interview with D. Trombardori, *Dits et écrits, II 1976–1988* (Paris: Gallimard, 2001), 861.
8. Claude Levi-Strauss, 'Humanity, what is it? An interview with Claude Lévi-Strauss', *The Kroeber Anthropological Papers*, 35 (1966): 48. '... ethnology has always been charged with the task of exploring to the very limits that which is under consideration at a given moment, in a given time, such as the limits of humanity ... We work for a science that ought always to grasp the fringes of the unknown (*La bordure de l'inconnu*). In that sense there will always be ethnology.'
9. *Ibid.*
10. Jean Rouch, 'Mettre en circulation des objets inquiétants', *Nouvelle Critique*, no. 82, March 1975.
11. It is not surprising that Rouch thought of himself as both a medium of cine-trance and a kind of hacker *avant la lettre*, having hijacked the very technology of cinema. In a memorable interview with Hanns Zischler, he declares that 'the future belongs to children and madmen', praises anthropology as a form of resistance to capitalism and rereads his own itinerary as an act of hacking, both of knowledge and of cinema. Hanns Zischler, *Jean Rouch, Ingenieur, Abenteurer, Ethnologe, Filmmacher* (1977, duration 40 min.).
12. On the notion of 'fringes', see William James's discussion of the fringes of the phenomenal in *Principles of Psychology*, vol. I (London: Macmillan and Co., 1891 [1890]), 258.
13. Henri Michaux, *Darkness Moves: An Henri Michaux Anthology*, 1927–1984, Selected, translated and presented by David Ball (Berkeley: University of California Press, 1994), 215.
14. Let's not think that there could be a criterion for defining an experiment with experience in the humanities. The experimental endeavour in the experimental sciences is endless, abyssal. It is simply that particular experiments submit to the constraint of producing results, hence arbitrary stopping points are attributed to them.
15. None would prevent one from first doing fieldwork in the classical sense of the term.
16. Henri Michaux, *Connaissance par les gouffres* (Paris: Gallimard, 1967), 232.
17. I would like to thank Graham Jones (MIT), who was the first to call Bappa *Bappa-Bobbamathing* (Bappa le Bidule), a true compliment coming from this anthropologist who specializes in magic, author of *Magic's Reason. An Anthropology of Analogy* (Chicago: Chicago University Press, 2017).
18. Regarding metaphysical debates provoked by the notion of artificial intelligence, see the classic work edited by Alan Ross Anderson, *Mind and Machines* (Englewood Cliffs, NJ: Prentice Hall, 1964).
19. Joffrey Becker provides a nice inventory of machines that have provoked never-ending enigmas, in his book *Humanoïdes, expérimentations croisées entre arts et sciences* (Nanterre: Presses Universitaires de Nanterre, 2015), from the Mechnical Turk to the Difference Engine (Babbage Machine), or else, from the Turing machine HAL in *2001 Space Odyssey*, all the way to the Ishiguro Geminoid, to give just a few examples.
20. George Marcus, 'Prototyping and contemporary anthropological experiments with ethnographic method', *Journal of Cultural Economy*, 7(4) (2014): 399–410.

21. Jean-Michel Truong makes this observation in his book *Totalement inhumaine* (Paris: Seuil, 2003). More recently, we could mention the coronavirus pandemic, which has only reinforced the massive digitalization of the economy and social relations and triggered a massive investment in tracking algorithms.
22. Gilles Deleuze, lecture on Foucault, 25 March 1986, https://www.webdeleuze.com/textes/281.
23. Lewis Mumford, *The Myth of the Machine, vol. I* (New York: Harcourt Brace Jovanovich, 1967); *The Myth of the Machine, vol. II* (New York: Harcourt Brace Jovanovich, 1970).
24. Let us quote, for example, the persecuting God of Major Schreber who sends him electric shocks, or the device to influence schizophrenics which controls the organs of Miss Natalia (Viktor Tausk's patient), to take up references dear to Deleuze and Guattari.
25. This is what he calls computers, after machine tools and automatons.
26. https://www.webdeleuze.com/textes/281. But one can refer to Gilles Deleuze, *Foucault*, trans. Sean Hand (Continuum, 2006 [1986]), 102 ('Appendix: On the Death of Man and Superman') to compare the two versions of the argument, oral and written.
27. Gilles Deleuze, 'Foucault', seminar at Université Paris 8, 8 April 1986, https://deleuze.cla.purdue.edu/seminars/foucault/lecture-18.
28. *Ibid.*
29. *Ibid.*
30. On Dick the prophet, mystic, but also anthropologist, see Pierre Déléage and Emmanuel Grimaud (eds), 'Estrangemental', *Gradhiva*, no. 29 (Paris: Quai Branly, 2019).
31. Philip K. Dick, 'The android and the human' (1972) in Robert Latham (ed.), *Science Fiction Criticism: An Anthology of Essential Writings* (London: Bloomsbury, 2017).
32. Eduardo Vivieros de Castro, 'Claude Lévi-Strauss vu par Viveiros de Castro', interview with Marc Kirsch, *La Lettre du Collège de France*, Hors-série 2 (2008): 34–35.
33. For a detailed version of the theory of the Technological Singularity, see my commentary on Vernor Vinge, Emmanuel Grimaud, 'Silicium Prophétie. La Singularité technologique selon Vernor Vinge', *Terrain*, no. 71 (2019): 140–57.
34. Jacques Ellul, *The New Demons*, trans. C. Edward Hopkin (New York: Seabury, 1975).
35. Wiener, *God & Golem, Inc.*.
36. Pierre Teilhard de Chardin, *The Human Phenomenon: A New Edition and Translation of Le phénomène humain*, trans. Sarah Appleton-Weber (Brighton: Sussex Academic Press, 2003 [orig. 1955]).
37. The expression *celestial computing* is taken from Mark Alizart. The philosopher suggests seeing Hegel as a thinker of computer science *avant la lettre* in his *Informatique celeste*.
38. After the idea of the Technological Singularity, Digital Gaia is one of the strongest hypotheses formulated by science fiction writer Vernor Vinge. Extrapolating from the growth curve of the internet, until it becomes a kind of ubiquitous sensory network connected to everything, he says that 'a web beneath the web' will emerge, one made up of all the networked

microprocessors sifting through the planet, including those attached to us and those that fall under the Internet of Things, forming 'distributed systems of fine-grained sensors'. This implemented animism, as he calls it, is one of the strangest possibilities envisaged by Vinge, as the world would then become 'its own database'. This electronic awakening of the Earth would, in his view, render our common sense of the world invalid, especially since one could well imagine this network entering into negotiations with old Gaia directly, without passing through any human mediator. See Vernor Vinge, 'The digital Gaia', *Wired*, 1 January (2000): 75, as well as James Lovelock, author of *The Gaia Hypothesis*, who takes up Vinge's scenario in *A Rough Ride to The Future* (London: Penguin, 2014).

39. 'The Singularity will happen when we go out and make it happen', writes Eliezer S. Yudkowsky, known at Berkeley for popularizing the idea of friendly artificial intelligence, in an online essay first published in 1996, at the age of 17, titled 'Staring into the singularity' (Berkeley, 1996), https://www.ondra.systems/source/staring-into-the-singularity-1-2-5.
40. That there is a link between the Turing test and the prophecies mentioned above seems obvious. Turing himself fuels them: 'At some stage therefore we should have to expect the machines to take control, in the way that is mentioned in Samuel Butler's '*Erewhon*'', he concludes in his article 'Intelligent machinery, a heretical theory', published posthumously in 1951, reprinted in *Philosophia Mathematica*, no. 4(3) (1996): 256–60.
41. We owe to the German philosopher and sociologist Helmut Plessner an original philosophy of ex-centricity, inspired by the ethology of Von Uexküll. According to him, the human being is always in an 'excentric' position in relation to its position in the world, either inside or outside itself. It is what pushes Plessner to write that 'An excentric being without equilibrium, standing out of place and time in nothingness, constitutively homeless, he must "become something" and create his own equilibrium.' Helmut Plessner, *Levels of Organic Life and the Human: An Introduction to Philosophical Anthropology*, trans. Millay Hyatt (New York: Fordham University Press, 2019 [orig. 1928]).
42. *L'homme hors de lui* (Paris: POL Editeur, 2018) is a beautiful play by playwright Valère Novarina.
43. On this point see, Timothy Morton, *Hyperobjects: Philosophy and Ecology after the End of the World* (Minneapolis: University of Minnesota Press, 2013). On the notion of 'reset' see the exhibition catalogue, Bruno Latour (ed.), *Reset Modernity!* (Boston: MIT Press, 2016).
44. At the time of *Ganesh Yourself*, transhumanists were not the only ones who were convinced that, since humanity was doomed, it was necessary to accelerate things and bring about the Singularity as quickly as possible. Accelerationists applied the same reasoning to the world economy, advocating for a collapse in order to get out of capitalism faster. On the curious ideas that come to humans as the prospect of their own demise becomes clearer, see Viveiros de Castro and Déborah Danowski, 'L'Arrêt de monde', in *De l'Univers clos au monde infini*, ed. Émilie Hache (Bellevaux: Éditions Dehors, 2014), as well as the special issue of the journal *Terrain*: 'Apocalypses', *Terrain*, no. 71 (2019).

45. Terence Deacon, *Incomplete Nature: How Mind Emerged from Matter* (New York: W. W. Norton & Company, 2011).
46. The text that deals with *homo metaphysicus*'s delirium in trading with the Infinite, continues thus: 'Exiled in the Infinite, deprived of the finite, in an incessant failure of his finite, he is no longer on the surface but in the center, a pure center at once and incredibly carried off course, as if at the focal point of multiple ellipses.' Henri Michaux, *Connaissance par les gouffres* (Paris: Gallimard, 1967), 232.
47. 'FRAGMENTATION: This is *indispensable* if one does not want to fall into REPRESENTATION. To see beings and things in their separate parts. To isolate these parts. Render them independent in order to give them a new dependence.' Robert Bresson *Notes on Cinematography* (Urizen, 1977 [orig. 1975]), 46.

Ganesh Yourself, the Film in Images

The experience recounted in this book gave rise to a film entitled *Ganesh Yourself* (Arte/Rouge International, 63 min., 2016) directed by Emmanuel Grimaud and produced by Nadia Turincev. Before Zaven Paré gave the interface its final form, as a remote-controlled device, many drawings were made. The photographs were taken during filming.

Figure 12.1. 'What monstrous forms could the absolute take on as our world continues to evolve?'

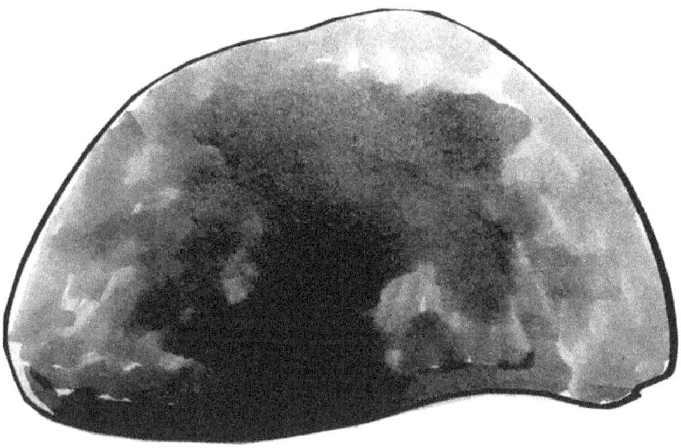

Figure 12.2. 'At that moment, Karan had only one desire, to escape, to dissolve, to return into the state of stone, to take refuge in deep meditation, the only viable state for the god he wanted to be. And he had the bright idea of sketching out a retreat by clinging to the only formula that came to his mind at that moment: "*Oṃ namaḥ śivāyam ... Oṃ namaḥ śivāyam ... Oṃ namaḥ śivāyam ...*"'

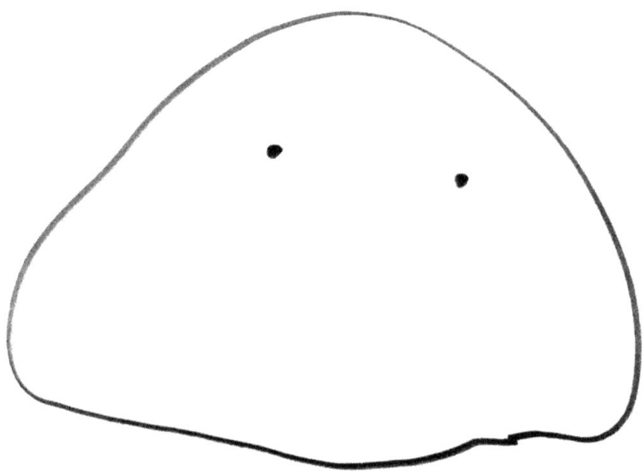

Figure 12.3. 'Self-generated (*svayambhū*) is he who comes and goes of his own will. Not *svayambhū* is he who is born in his mother's uterus; not *svayambhū* is he who is afraid of death. *Svayambhū* is he who is neither born nor dies,' the *paṇḍit* replied. 'He comes when you need him, and leaves when your need vanishes. But when you invoke him, he comes ...'

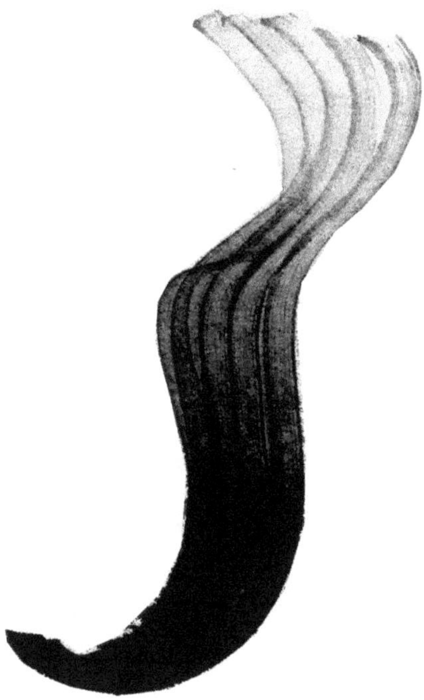

Figure 12.4. 'Equipped with a thermoformed mask and a trunk, no doubt a bit abstract, it was elephantine all the same. There was no mistaking it, the machine had something a bit Ganesh about it.'

Figure 12.5. 'The system, tele-operated by a human, which we called an "incarnant", and in which the only truly robotic component is, in fact, his trunk, animated by sensors, is at first glance quite close to Philip K. Dick's "walkie-talkie", the idea of a portable god or system that allows a divine entity to teleport anywhere.'

Figure 12.6. 'It would be a mistake to see here only an experiment in implementing a god machine, an experiment whose aim would be to evaluate its conditions of viability, as if machines, ever more pervasive, were not already the object of a cult.'

Figure 12.7. 'This seemingly divine gadget or simulacrum of the god Ganesh is in fact a machine of divine deconstruction and might prefigure other forms, another species, another pact with animals and machines.'

Figure 12.8. 'Don't be shy. I have come here for you. I have come from far away, from Mount Kailash, just for my devotees. If you do not come to me, I will disappear forever.'

Figure 12.9. 'No one could say, before doing the experiment, whether a little more tinkering, a bit of electricity, artificial intelligence and a tele-operated voice, would or would not generate something else, something unknown, something not entirely assimilable to the paths of the sacred already encountered.'

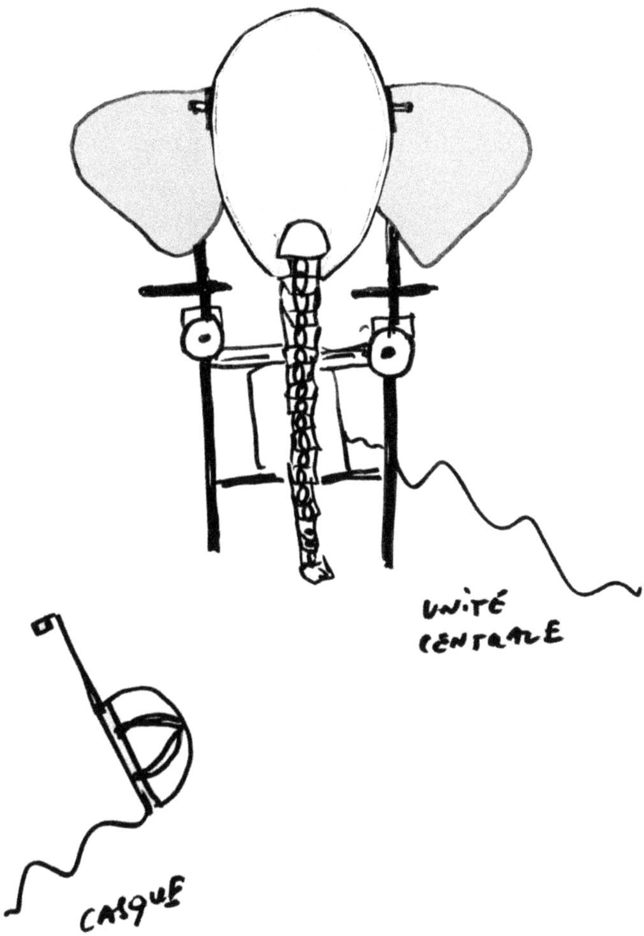

Figure 12.10. 'It is there, in the hollows of a process of incarnation that we thought illusory or doomed to failure, that the multitude of possible gods will be best felt, including those for which we have not opted, those we could have adopted or those we could make our own.'

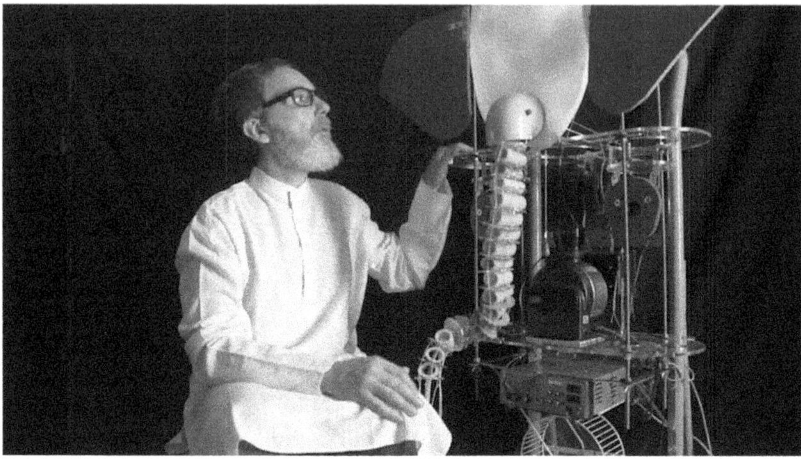

Figure 12.11. Zaven Paré testing the machine. The garland visible on the machine shows that it has undergone an eye-opening ritual, as with any other idol. 'A garland was placed on it, and on its forehead the officiant invited us to apply powder using the thumb at the locus of the third eye (*ājñā*), in order to open its "subtle channel".'

Figure 12.12. Emmanuel Grimaud and Zaven Paré calibrating the device.

Figure 12.13. Karan, the first incarnation, prepares to face questions near a Ganesh altar designed for Ganesa Chaturthi.

Figure 12.14. In its tele-operated form, the new variant of the God, a singular mix between a voice of proximity and an image at a distance, allowed repressed reproaches to be said. 'Are our problems hidden from you? If you are God, don't you know our wishes and desires?'

Figure 12.15. The robot suddenly appears overwhelmed by a crowd of questioners. 'Waves of interaction followed suit, the one after the other. Tests followed by showers of insults, followed by new waves of cross-examination for the attainment of divine status.'

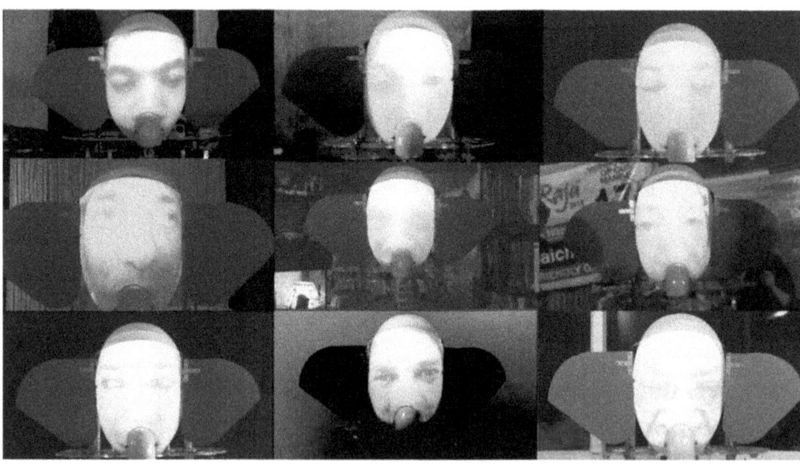

Figure 12.16. Some incarnations: Karan (top left), Varsha (top centre), Jyot the transgender Ganesh (top right), Anand P. the environmentalist Ganesh (2nd row left), Harish the LGBT Ganesh (2nd row centre), Anand T. the atheist Ganesh (2nd row right), Jyot the transgender Ganesh (3rd row left), Modi the astrologer Ganesh (3rd row centre), Pr Kumar, the anti-radiation Ganesh (3rd row right).

Figure 12.17. 'This time Anand did not answer. Tired of not being able to have a real dialogue in the way that he understood it, he would let the other person take up the deep reason for his existence, making public the very emptiness he was feeling at that moment.'

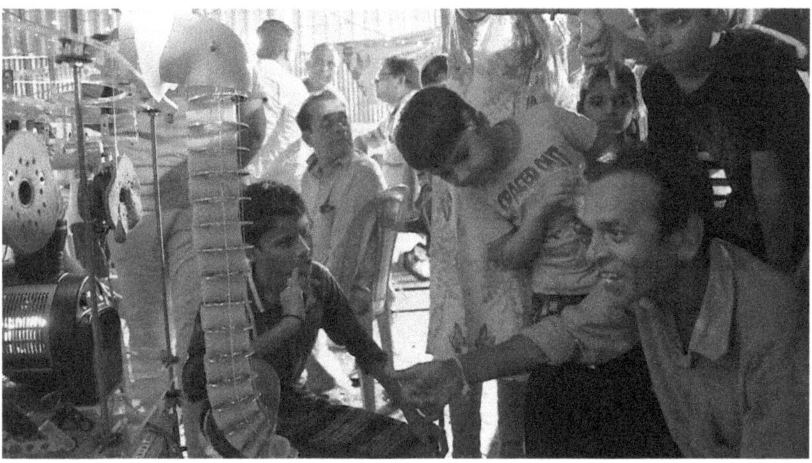

Figure 12.18. 'Divine ventriloquism and, as with all ventriloquism, "the opportunity for strange voices to express themselves", which are not reducible either to the ventriloquist or to the one he manipulates.'

Figure 12.19. 'Spectrosymbiosis: the machine (host entity) and the god (parasitic entity) exchange links, with the complicity of the tele-operator (the incarnant) who is also someone's shadow, the ghost of something, a hacker or pirate of God, who is in turn hackable by God or by other voices.'

Figure 12.20. 'An unstable spectral glow, by way of an overhead projector, as fragile as it is capricious.'

BIBLIOGRAPHY

Alizart, Mark, *Informatique celeste*, Paris: Presses universitaires de France, 2018.

Allen Hynek, J., *The UFO Experience: A Scientific Inquiry*, Washington: H. Regnery Company, 1972.

Anderson, Alan Ross (ed.), *Mind and Machines*, Englewood Cliffs, NJ: Prentice Hall, 1964.

Assman, Jan, *Moses the Egyptian: The Memory of Egypt in Western Monotheism*, Boston: Harvard University Press, 1998.

Assman, Jan, *The Price of Monotheism*, trans., Robert Savage, Stanford: Stanford University Press, 2010.

Baudouin, Philippe, 'Archéologie des machines occultes', *Terrain*, no. 69 (2018): 96–113.

Becker, Joffrey, *Humanoïdes, expérimentations croisées entre arts et sciences*, Nanterre: Presses Universitaires de Nanterre, 2015.

Bénard, Charles, *Sur l'esthétique de Hegel*, Paris: Ladrange, 1852.

Bernard, Claude, *Principes de médecine expérimentale*, Paris: Presses universitaires de France, 2008 [1877].

Berti, Daniela, *La Parole des Dieux*, Paris: CNRS, 2001.

Bettini, Maurizio, *Éloge du polythéisme*, Paris: Les Belles Lettres, 2016.

Bouillier, Véronique and Gérard Toffin (eds), *Classer les dieux?* Collection Purushartha, vol. 15, Paris: EHESS, 1993.

Boyer, Pascal, *Religion Explained: The Human Instincts that Fashion Gods, Spirits and Ancestors*, New York: Basic Books, 2001.

Braitenberg, Valentino, *Vehicles: Experiments in Synthetic Psychology*, Cambridge, MA: MIT Press, 1984.

Bresson, Robert, *Notes on Cinematography*, Urizen, 1977 [orig. 1975].

Butler, Samuel, *Erewhon or Over the Range*, London: Trübner and Co., 1872.

Capek, Karel, *The Absolute at Large*, trans. Šarká B. Hrbková, London: Macmillan, 1927 [orig. 1922].

Carey, Matthew, 'What is the plural of Apocalypse? An essay on the ends of time', *Terrain* [online], no. 71 (2019).

Cassin, Barbara and Mark Alizart (eds), *Traces du sacré*, Paris: Centre Georges Pompidou, 2008.

Coleridge, Samuel Taylor, *Biographia Literaria*, 1817.

Cooren, François, *Manières de faire parler. Interaction et ventriloquie*, Paris: Le Bord de l'eau, 2013.

Criado, Tomás Sánchez and Adolfo Estalella, 'The ethnographic invention', in *An Ethnographic Inventory: Field Devices for Anthropological Inquiry*, ed. Tomás Sánchez Criado and Adolfo Estalella, London: Routledge, 2023.

Davis, Erik, *TechGnosis. Myth, Magic and Mysticism in the Age of Information*, London: Serpent's Tail, 1999.

Deacon, Terence, *Incomplete Nature: How Mind Emerged from Matter*, New York: W. W. Norton & Company, 2011.

de Castro, Eduardo Viveiros, 'Claude Lévi-Strauss vu par Viveiros de Castro', interview with Marc Kirsch, *La Lettre du Collège de France*, Hors-série 2 (2008): 34–35.

de Castro, Eduardo Viveiros and Déborah Danowski, 'L'Arrêt de monde', in *De l'Univers clos au monde infini*, ed. Émilie Hache, Bellevaux: Éditions Dehors, 2014.

de Chardin, Pierre Teilhard, *The Human Phenomenon: A New Edition and Translation of Le phénomène humain*, trans. Sarah Appleton-Weber, Brighton: Sussex Academic Press, 2003 [orig. 1955].

Déléage, Pierre and Emmanuel Grimaud (eds), Estrangemental, *Gradhiva*, no. 29, Paris: Quai Branly, 2019.

Deleuze, Gilles, lecture on Foucault, 25 March 1986, https://www.webdeleuze.com/textes/281.

Deleuze, Gilles, 'Foucault', Seminar at Paris 8 University, 8 April 1986, https://deleuze.cla.purdue.edu/seminars/foucault/lecture-18, translation modified.

Deleuze, Gilles, *Foucault*, 'Appendix: On the Death of Man and Superman', trans. Sean Hand, London: Continuum, 2006 [1986].

de Samosate, Lucien, *Alexandre ou le faux prophète*, vol. I, Paris: Hachette, 2002.

Descola, Philippe, 'La forêt des signes', Preface in Eduardo Kohn, *How Forests Think. Comment pensent les forêts*, Brussels: Zones sensibles, 2017, 11–17.

Despret, Vinciane, *Hans le cheval qui savait compter*, Paris: Les Empêcheurs de penser en rond, 2004.

Despret, Vinciane, *Habiter en oiseau*, Paris: Actes Sud, 2019.

Dewey, John, *Experience and Nature*, Chicago: Open Court, 1925.

Dewey, John, *Art as Experience*, New York: Perigee, 1980 [1934].

Dick, Philip K., *The Exegesis of Philip K. Dick*, ed. Pamela Jackson and Jonathan Lethem, New York: Houghton Mifflin Harcourt, 2011.

Dick, Philip K., 'The android and the human' (1972) in Robert Latham (ed.), *Science Fiction Criticism: An Anthology of Essential Writings*, London: Bloomsbury, 2017.

Dorthe, Gabriel, 'L'Immortalité et ses impatients', *Socio-anthropologie*, no. 31 (2015): 127–38.

Dufrene, Thierry, Emmanuel Grimaud, Denis Vidal and Anne-Christine Taylor (eds), *Persona, étrangement humain*, Paris: Musée du quai Branly/ Actes Sud, 2016.

Dunne, Anthony and Fiona Raby, *Speculative Everything: Design, Fiction and Social Dreaming*, Boston: MIT Press, 2013.

Ellul, Jacques, *The New Demons*, trans. C. Edward Hopkin, New York: Seabury, 1975.

Estalella, Adolfo and Tomás Sánchez Criado (eds), *Experimental Collaborations. Ethnography Through Fieldwork Devices*, New York, Oxford: Berghahn, 2018.

Farman, Abou, 'Re-enchantment cosmologies: Mastery and obsolescence in an intelligent universe', *Anthropological Quarterly*, no. 4 (2012): 1069–88.

Fischer, Michael, *Anthropology in the Meantime: Experimental Ethnography, Theory, and Method for the Twenty-First Century*, Durham: Duke University Press, 2018.

Foucault, Michel, 'Entretien avec Michel Foucault', interview with D. Trombardori, *Dits et écrits, II 1976–1988*, Paris: Gallimard, 2001.

Gell, Alfred, 'Technology and magic', *Anthropology Today*, no. 4(2) (1988): 6–9.

Gell, Alfred, 'Vogel's net: Traps as artworks and artworks as traps', *Journal of Material Culture*, no. 1 (1996): 15–38.

Goldblatt, David, *Art and Ventriloquism*, London: Routledge, 2006.

Griaule, Marcel, 'Cours sur les techniques' (11 March 1953), in *Méthode de l'ethnographie*, Paris: Presses universitaires de France, 1957.

Grimaud, Emmanuel, *Dieux et Robots*, Apt: L'Archange Minotaure, 2008.

Grimaud, Emmanuel, 'Pour quelques secondes de confusion ontologique. Échecs et réussites du contact dans une expérience de télé-robotique', in Thierry Dufrene and Joël Huthwohl (eds), *La Marionnette*, Montpellier: L'Entretemps, 2014.

Grimaud, Emmanuel, *L'étrange Encyclopédie du docteur K.*, Nanterre: Société d'ethnologie, 2014.

Grimaud, Emmanuel, 'Silicium Prophétie. La Singularité technologique selon Vernor Vinge', *Terrain*, no. 71 (2019): 140–57.

Grimaud, Emmanuel and Zaven Paré, *Le Jour où les robots mangeront des pommes. Conversations avec un Geminoid*, Paris: Petra, 2012.

Grimaud, Emmanuel and Anthony Stavrianakis, 'We have never been not experimental/how much more experimental can anthropology be?', *Annual Review of Anthropology*, forthcoming.

Grimaud, Emmanuel, Yann-Philippe Tastevin and Denis Vidal (eds), *Low Tech? Wild Tech!* Techniques et Culture 67, 2017.

Grimaud, Emmanuel, Anne-Christine Taylor, Denis Vidal and Thierry Dufrêne, 'Qui est là? Présences limites et effets de personne', in *Persona, étrangement humain*, Paris: Musée du quai Branly/Actes Sud, 2016, 11–19.

Guenzi, Caterina, *Le discours du destin*, Paris: CNRS, 2013.

Hegel, G. W. F., trans. T. M. Knox, *Hegel's Aesthetics: Lectures on Fine Art*, Oxford: Clarendon Press, 1975.

Heider, Fritz and Marianne Simmel, 'An experimental study of apparent behavior', *American Journal of Psychology*, vol. 57 (1944): 243–59.

Herrenschmidt, Olivier, 'Choisir ses dieux', in *Persona, étrangement humain*, Paris: Actes Sud, 2016, 214–15.

Hulin, Michel, *La Mystique sauvage, aux antipodes de l'esprit*, Paris: Presses universitaires de France, 1993.

Hulin, Michel, *Comment la philosophie indienne s'est développée: la querelle brahmanes-bouddhistes*, Paris: Éditions du Panama, 2008.

James, William, *Principles of Psychology*, vol. I, London: Macmillan and Co., 1891 [1890].

James, William, *The Will to Believe and Other Essays in Popular Philosophy*, New York: Longmans, Green and Co., 1897.

Jones, Graham, *Magic's Reason: An Anthropology of Analogy*, Chicago: Chicago University Press, 2017.

Kaur, Raminder, *Performative Politics and the Cultures of Hinduism: Public Uses of Religion in Western India*, London: Anthem South Asian Studies, 2003.

Koch, Renée, *Comment peut-on être dieu, la secte d'Épicure*, Paris: Belin, 2005.

Kurzweil, Ray, *The Singularity is Near. When Humans Transcend Biology*, New York: Viking Books, 2005.

Lafitte, Jacques, *Réflexions sur la science des machines*, Paris: Vrin, 1972 [1932].

Laing, Ronald D., *Knots*, London: Tavistock Publications, 1970.

Latour, Bruno (ed.), *Reset Modernity!* Boston: MIT Press, 2016.

Leroi-Gourhan, André, *Milieu et Techniques*, Paris: Albin Michel, 1973.

Levi-Strauss, Claude, 'Humanity, what is it? An interview with Claude Lévi-Strauss', *The Kroeber Anthropological Papers*, 35 (1966): 41–53.

Lovelock, James, *A Rough Ride to The Future*, London: Penguin, 2014.

Malamoud, Charles, *Cooking the World: Ritual and Thought in Ancient India*, trans. David White, Oxford: Oxford University Press, 1996 [1989].

Malamoud, Charles, *La Danse des pierres. Études sur la scène sacrificielle dans l'Inde ancienne*, Paris: Seuil, 2005.

Marcus, George, 'Prototyping and contemporary anthropological experiments with ethnographic method', *Journal of Cultural Economy*, 7(4) (2014): 399–410.

Margulis, Lynn and Dorion Sagan, *Microcosmos: Four Billion Years of Evolution from Our Microbial Ancestors*, Berkeley: University of California Press, 1986.

Mauméjean, Xavier, *Ganesha, Mémoires de l'homme-éléphant*, Paris: Mnemos Editions, 2007.

Michael, S. M., 'The origin of the Ganapati Cult', *Asian Folklore Studies*, no. 42(1) (1983): 91–16.

Michaux, Henri, *Darkness Moves: An Henri Michaux Anthology*, 1927–1984, Selected, translated and presented by David Ball, Berkeley: University of California Press, 1994.

Michaux, Henri, *Connaissance par les gouffres*, Paris: Gallimard, 1967.

More, Max, 'Principles of Extropy (Version 3.11): An evolving framework of values and standards for continuously improving the human condition', Extropy Institute, 2003.

Mori, Masahiro, *The Buddha in the Robot: A Robot Engineer's Thoughts on Science and Religion*, Tokyo: Rissho Kosei-kai, 1981.

Morton, Timothy, *Hyperobjects: Philosophy and Ecology after the End of the World*, Minneapolis: University of Minnesota Press, 2013.

Mumford, Lewis, *The Myth of the Machine, vol. I*, New York: Harcourt Brace Jovanovich, 1967.

Mumford, Lewis, *The Myth of the Machine, vol. II*, New York: Harcourt Brace Jovanovich, 1970.

Murphy, Keith and Eitan Y. Wilf (eds), *Designs and Anthropologies: Frictions and Affinities*, Santa Fe: School for Advanced Research Press, 2021.

Novarina, Valère, *L'homme hors de lui*, Paris: POL Editeur, 2018.

O'Connell, Mark, *To Be A Machine: Adventures Among Cyborgs, Utopians, Hackers, and the Futurists Solving the Modest Problem of Death*, New York: Doubleday, 2017.

Paré, Zaven, Chihiro Minato and Emmanuel Grimaud, 'Le Bouddha dans le robot', *Gradhiva*, no. 15 (2012): 162–81.

Plessner, Helmut, *Levels of Organic Life and the Human: An Introduction to Philosophical Anthropology*, trans. Millay Hyatt, New York: Fordham University Press, 2019 [orig. 1928].

Rabinow, Paul and Gaymon Bennett, *Designing Human Practices: An Experiment With Synthetic Biology*, Chicago: University of Chicago Press, 2012.

Ray, Satyajit, *The Goddess*, film, India, 93', 1960.

Rheinberger, Hans-Jorg, *Toward a History of Epistemic Things: Synthesizing Proteins in the Test Tube*, Stanford, CA: Stanford University Press, 1997.

Rouch, Jean, 'Mettre en circulation des objets inquiétants', *Nouvelle Critique*, no. 82 (March 1975).

Seaver, Nick, 'Captivating algorithms: Recommender systems as traps', *Journal of Material Culture*, no. 24(4) (2019): 421–36.

Simondon, Georges, *Du mode d'existence des objets techniques*, Paris: Aubier, 1958.

Soldhju, Katrin, *Selbstexperimente. Die Suche nach der Innenperspektive und ihre epistemologischen Folgen*, Berlin: Trajekte, 2012.

Truong, Jean-Michel, *Totalement inhumaine*, Paris: Seuil, 2003.

Tucker, Mary Evelyn and John A. Grim (eds), *Worldviews and Ecology*, New York: Orbis Books, 1994.

Turing, Alan Mathison, 'Computing machinery and intelligence', *Mind*, no. 59 (1950): 433–60.

Turing, Alan Mathison, 'Intelligent machinery, a heretical theory', published posthumously in 1951, reprinted in *Philosophia Mathematica*, no. 4(3) (1996): 256–60.

Vidal, Denis, *Le Culte des divinités locales dans une région d'Himachal Pradesh*, Paris: ORSTOM, 1988.
Vidal, Denis, *Frontières de l'humain. Dieux, figures de cire, robots, et autres artefacts*, Paris: Alma, 2016.
Vinge, Vernor, 'The coming Technological Singularity: How to survive in the post-human era', *Whole Earth Review*, 1993.
Vinge, Vernor, 'The digital Gaia', *Wired*, 1 January (2000): 75.
White, Lynn Townsend Jr, 'The historical roots of our ecological crisis', *Science*, no. 155 (1967): 1203–7.
Whitehead, Alfred North, *Religion in the Making: Lowell Lectures*, Cambridge: Cambridge University Press, 1926.
Whitehead, Alfred North, *Process and Reality; An Essay in Cosmology. Gifford Lectures 1927–1928*, corrected edition, ed. D. R. Griffin and D. W. Sherburne, New York: The Free Press, [1978] 1985.
Wiener, Norbert, *God & Golem, Inc. A Comment on Certain Points Where Cybernetics Impinges on Religion*, Cambridge, MA: MIT Press, 1964.
Wissowa, Georg, *Religion und Kultus der Römer*, Munich: Beck, 1912.
Yudkowsky, Eliezer S., 'Staring into the singularity', online essay, Berkeley, 1996, https://www.ondra.systems/source/staring-into-the-singularity-1-2-5.
Yudkowsky, Eliezer S., *The Singularitarian Principles*, Version 1.0.2, Compact edition, 2000.
Zischler, Hanns, *Jean Rouch, Ingenieur, Abenteurer, Ethnologe, Filmmacher*, 1977, duration 40 min.

INDEX

www.ingramcontent.com/pod-product-compliance
Ingram Content Group UK Ltd.
Pitfield, Milton Keynes, MK11 3LW, UK
UKHW021837150926
8656IPUK00004B/27

9 781836 955276